Mathematical Finance: A Very Short Introduction

VERY SHORT INTRODUCTIONS are for anyone wanting a stimulating and accessible way into a new subject. They are written by experts, and have been translated into more than 45 different languages.

The series began in 1995, and now covers a wide variety of topics in every discipline. The VSI library currently contains over 550 volumes—a Very Short Introduction to everything from Psychology and Philosophy of Science to American History and Relativity—and continues to grow in every subject area.

Very Short Introductions available now:

For more information visit our website
www.oup.com/vsi/

Mark H. A. Davis

MATHEMATICAL FINANCE

A Very Short Introduction

OXFORD
UNIVERSITY PRESS

OXFORD
UNIVERSITY PRESS

Great Clarendon Street, Oxford, OX2 6DP,
United Kingdom

Oxford University Press is a department of the University of Oxford.
It furthers the University's objective of excellence in research, scholarship,
and education by publishing worldwide. Oxford is a registered trade mark of
Oxford University Press in the UK and in certain other countries

First Edition published in 2019

Published in the United States of America by Oxford University Press
198 Madison Avenue, New York, NY 10016, United States of America

British Library Cataloguing in Publication Data
Data available

Library of Congress Control Number: 2018957177

ISBN 978-0-19-878794-5

Printed and bound by CPI Group (UK) Ltd, Croydon, CR0 4YY

Contents

Preface

Mathematical finance (MF) is a body of mathematical, statistical, and computational techniques that by now are ubiquitous in the world of finance for valuing and managing the risks of financial contracts and the mechanisms by which they are traded. Until recently MF was tightly focused on questions of 'arbitrage' (the availability of riskless profit) and on financial derivatives—contracts whose value is 'derived' from the value of some underlying asset such as a stock price or foreign exchange rate. Nowadays, in response to the financial crisis and to rapid developments in technology, the range of questions is much broader, with increased emphasis on the mechanisms by which prices are determined as well as new angles on, for example, algorithmic trading.

MF is rooted in five disciplines: financial economics, statistics, probability theory (or 'stochastic analysis'), numerical analysis, and computer technology. For a long time, and somewhat counter-intuitively, the latter three were the dominant ones. The subject got going as a specialized topic in financial economics, but as it developed the mathematical aspects took over as they proved to be the relevant ones for the study of financial options. Statistics were always used, but in quite conventional ways. The use of sophisticated statistical methods has massively increased in recent years, and financial economics has enjoyed a modest resurgence in

MF, while the emphasis on the traditional topics of option pricing and hedging has somewhat declined; the reasons for these trends are the subject of the final Chapter 8.

Today's MF began with the systematic study of option valuation initiated by economist Paul Samuelson and others at MIT in the 1960s. Samuelson, who was awarded the economics Nobel Prize in 1970 for 'raising the level of analysis in economic science', had picked up on much earlier work by Louis Bachelier (details in Chapter 3) and realized that stochastic analysis was the key 'tool' that he needed. 'Stochastic' is just another word for 'random', and use of the term 'stochastic processes' to denote the mathematical study of random motions such as stock prices or weather patterns was set in concrete by J.L. Doob's highly regarded 1953 textbook of that title. Of particular interest to Samuelson was the 'stochastic calculus' introduced by Japanese mathematician Kiyosi Itô. It is a curious fact that the mathematical theories relevant to finance came from the purest of pure mathematics rather than what we normally think of as applied mathematics. In the early 1960s only a few *aficionados* had ever heard of Itô calculus. Now everybody who has taken a Master's course in MF knows all about it.

The option pricing problem was finally cracked by Fischer Black and Myron Scholes (1973), who derived the Black–Scholes formula, the most famous formula in financial economics. An independent derivation in much the form found in modern textbooks was given in the same year by Robert Merton of the MIT group. The result led to an explosion of activity developing the theory and trading in the markets.

One by-product of this expansion was that, increasingly, PhD scientists (mostly physicists at first) were sucked into the markets forming a breed of 'quants' (quantitative analysts) who were able to handle the technical complexities of the subject. At the same time, there were rapid advances in computer technology, without which options markets could not have functioned. Over time, universities started offering specialist courses in this area, but it is still true that students and professionals from a wide range of science and engineering disciplines find their way into the markets.

Turning to the contents of the book, Chapter 1 describes the playing field: the world of money, banking, and financial markets. It outlines the assets that are traded as well as some of the trading mechanisms such as futures markets, and basic definitions of options, with some standard terminology. Without this information, you will not understand the purpose of the theory. Chapter 2 covers probability theory and its *frequentist* and *subjectivist* interpretations. The latter is included because of its close connection with the so-called Fundamental Theorem of Asset Pricing (FTAP), the single most important and far-reaching result in 'classical' MF, which is covered later, in Chapter 3. The reader is not expected to have previous knowledge of the topics in Chapter 2 beyond a familiarity with basic ideas in statistics (mean, standard deviation, the normal distribution . . .), standard mathematical items such as logarithms and exponentials, and a little calculus. Much of the content is accessible without all of this background.

It is beyond the scope of the book to attempt an explanation of Itô calculus, but, also in Chapter 3, the reader is introduced, in an intuitive manner, to Brownian Motion and to Einstein's explanation of why, for Brownian Motion, the ordinary rules of calculus cannot work. Our approach to the Black–Scholes formula is via the so-called binomial tree, for which there is a self-contained option pricing theory. This topic is covered in detail, and it is worth the effort of following these details because the concepts are there in a framework that is easily grasped. What comes later is arguably just 'technicalities' covering more complex cases but using the same ideas. In addition, the binomial tree can be interpreted as a computational method for calculating the Black–Scholes formula, which provides us with a justification for the latter.

Option pricing in its simplest form concerns options on assets such as stocks, where the price is just one number that moves over time. Equally important are options on interest-rate-related contracts, where the underlying asset value at any one time is not just a number but a whole curve, the so-called *yield curve*. This is the much more complicated, but arguably much more interesting, problem that is the subject of Chapter 4. The next chapter broaches the subject of *credit risk*, the risk that your counterparty might default on future obligations. Trading in credit risk increased from next to nothing to a huge volume in the early part of this century, and matters related to it were in no small part to blame for the financial crisis of 2008.

Chapter 6 moves on to discuss an apparently very different subject, namely fund management where the objective is to form portfolios of assets so as to maximize the investment return. Here, an MF-oriented approach was kicked off by Robert Merton in 1969. Fund management is a huge industry—and has become much more technical with the emergence of hedge funds deploying sophisticated strategies—so such studies are important, even though the impact of theory on practice in this area has been much less than in the topics covered in previous chapters. There are also new insights, uncovering features of the stock market that had hitherto passed by unnoticed.

In every finance house there will be a trading floor, where traders do deals and, on the next floor up or down, a 'risk management' department whose job it is to monitor the trades and make sure the house is not taking on too much risk, and is complying with financial regulations. All financial institutions operate within a legal framework which imposes regulatory requirements. The rules have expanded in scope and tightened considerably in the post-crisis era, leading to a large increase in the risk management function and consequent deployment of quant effort. All of this is the subject of Chapter 7. This is an area where statistics plays the primary role.

The final chapter concerns the 2008 crisis and its aftermath. The year 2008 was a watershed moment for finance in many ways. For MF it has meant a complete re-orientation. The traditional topics of pricing and arbitrage theory live on and have not declined in importance. However, they are no longer the sole focus of research since fundamentally the theory is complete, and also because most of the more 'exotic' contracts have become unviable in the new regulatory environment. On the other hand, a whole raft of new questions concerning the behaviour and interaction of markets have opened up, and the implications of new technology, now generally known as FinTech (financial technology), must be taken on board.

Largely omitted from this book is discussion of numerical analysis and computation. This may seem absurd when the quants spend 80 per cent of their time developing algorithms and writing code, but it is impossible to do the topic justice within the confines of a VSI. And arguably much of it is generic, covering a much wider range of applications in science and engineering than just MF. I have preferred to concentrate on the conceptual content of topics that are specific to MF. Algorithms deserve a volume of their own.

I have learnt so many things from far too many people to acknowledge adequately here. But specifically I would like to mention Eugene Wong, from whom I took my first course in stochastic processes at Berkeley (how would I know where that was going to lead?) and Johan Beumée who taught me how to be a quant at Mitsubishi Finance. Elena Medova and Michael Dempster have been companions for much of the way, constant providers of ideas, inspiration, and encouragement. Finally, thanks as ever to my wife Jessica for her support and forbearance as I am tied up yet again in a writing project. At least this was a Very Short one.

Mark Davis
London, September 2018

List of illustrations

List of abbreviations

ALM	asset/liability management
ATM	at-the-money
BET	Moody's Binomial Expansion Technique
BGM	Brace, Gątarek, and Musiela model
BIS	Bank for International Settlements
bn	billion
BS	Black and Scholes
CDO	Collateralized Debt Obligations
CDS	credit default swaps
CIR	Cox–Ingersoll–Ross model
CLT	Central Limit Theorem
CVA	Credit Value Adjustments
DBA	Dutch Book arguments
DBT	Dutch Book Theorem
EMS	expected median shortfall
ES	Expected Shortfall
EUR	Euro, €
Fed	Federal Reserve Bank, New York
FRA	forward rate agreement
FTAP	Fundamental Theorem of Asset Pricing

FTSE100	total market capitalization of the biggest hundred firms traded on the London Stock Exchange (scaled)
FX	foreign exchange
GARCH	A commonly used econometric model
GBP	Pound Sterling, £
HJM	Heath, Jarrow, and Morton model
ICE	Intercontinental Exchange
i.i.d.	independent, identically distributed
IPO	initial public offering
LIBOR	London Interbank Offered Rate
LMM	Libor Market Model
LOB	limit order book
m	million
MIT	Massachusetts Institute of Technology
OTC	over-the-counter
PCA	Principal Components Analysis
PIT	probability integral transform
RMBS	residential mortgage-backed security
S&P	Standard and Poor's rating agency
S&P500	Standard and Poor's index of the biggest 500 firms traded on the New York Stock Exchange
SPV	special purpose vehicle
tn	trillion
USD	United States Dollar, $
VaR	Value at Risk
XVA	value adjustments (X being variable, dependent on VA being considered)
ZCB	zero coupon bond

Chapter 1
Money, banking, and financial markets

Money

Finance is about money and, when you think about it, money is an elusive concept. It is even more so in the era of contactless cards and mobile phone payment systems where money ceases to have any physical instantiation, and there is unresolved debate whether cryptocurrency payment systems such as Bitcoin qualify as money at all. In the commercial world, money is most of the time just numbers on a spreadsheet, and often these numbers are hard to relate to, being quoted in enormous units, millions (1m = 1,000,000), billions (1bn = 1,000m), or even trillions (1tn = 1,000bn). It is not uncommon to find monetary amounts in billions erroneously quoted in press reports as millions; that one-character typographical error makes quite a difference.

The need for a *medium of exchange* has been recognized since the earliest times. The original medium was *commodity money*, in which a fixed quantity of some specified commodity such as rice, olive oil, or copper is used as a currency unit, in terms of which prices of other things—be they other commodities, manufactured goods, or labour rates—are determined. A major conceptual and practical advance was the introduction of *fiat money* in which the value of any commodity is expressed in abstract units, say dollars, so that the exchange rate between any two commodities is simply the ratio of their dollar prices. To turn this idea into reality individuals must be able to store and exchange the dollars themselves, which was achieved by the introduction of coinage. Fiat money is built on trust: purchasers must be sure that if an item is priced at $10 and they hand over $10-worth of money they will receive the goods, and conversely the sellers must be sure that the money they receive is the real thing, not some home-produced fake. Trust can only be achieved by having some monetary authority, generally an arm of government, that issues currency, regulates the amount in circulation, and acts to limit counterfeiting. There are of course many instances in history where trust has broken down. Initially made from precious metals, coinage has evolved over time to a point where the physical value of the coin is a small fraction of its monetary value, so in practical terms it is equivalent to paper money, which was introduced in China during the Song Dynasty, AD 960–1127. (The West had hardly discovered paper, let alone paper money, at the time). It took a surprisingly long time for paper money to evolve into the pure fiat money we know today: right up to the First World War and beyond paper money was generally backed by gold, for which it could in theory be exchanged. Indeed, UK banknotes are still signed by the Chief Cashier of the Bank of England under the statement 'I promise to pay the bearer on demand the sum of £10'. Understandably, this commitment appears in a smaller and smaller font with each successive redesign of the note.

In addition to its role as a medium of exchange, money acts as a *store of value*. We get it today, but we don't have to spend it today. You could store it in a safe at home—an increasingly popular option in Japan after decades of near-zero interest rates—or you can open a current account at a bank. In the latter case it is clear that the balance in your current account is money no less than cash itself. The total value of current account deposits greatly exceeds the value of cash in circulation. In econometrics, 'narrow money' M_1 denotes the total value of cash and current account deposits. Its value in the UK in August 2016 was £1.62tn, of which £67.8bn was banknotes and a further £4bn was coins, implying that the value of bank deposits is more than twenty times that of notes and coins, a number that can only increase as electronic payments take over. For comparison, the gross domestic product (GDP) of the UK in 2015 was about £1.9tn. The *transactions demand for money* is the amount required to underpin the current rate of transactions taking place in the economy as a whole, and the central bank will supply sufficient notes and coins to support that fraction of the transactions that involve cash.

So far we have discussed money only in its role in day-to-day transactions. However, the longer term picture is at least as important, if not more so. As John Maynard Keynes put it 'the importance of money essentially flows from its being a link between the present and the future'. This link is formed in various ways by the activities of banks and financial markets.

On the simplest level, a bank operates by taking in deposits from account holders and making a profit by making loans, backed by the deposits, to individuals or companies, for which the bank charges interest. It was realized long ago that it is not necessary to limit the value of the loans to the value of the deposits; a 'reserve ratio' of say 10 per cent will provide adequate liquidity, that is, ability to service the day-to-day needs of account holders. So starting with £1m the bank could make loans of £9m. The recipients of the loans will, in the first instance, pay them into their current accounts, at which point the loans become money and contribute to M_1. In this way the 'fractional reserve' system hugely increases the money supply. The money supply is reduced again when the loan is repaid, but taking the economy as a whole there will always be a large number of outstanding loans. Bank loans are the simplest way in which resources can be brought back from the future to be used now, for diverse purposes such as developing a new product, buying a house, or paying student fees. In all cases the borrower's future resources are reduced by the

3

amount needed to pay off the loan and the lender has taken on the risk that the borrower may default.

Central banks do not directly control the money supply but have various indirect ways to influence the activities of the banks, all of which are regulated by the central bank in the jurisdiction in which they are operating. For example, the Bank of England sets a 'base rate', the rate at which it will lend money to commercial banks. This influences the rates at which those banks can lend to customers, which in turn affects the amount of lending that will occur.

Financial markets

The financial markets as a whole provide a myriad of mechanisms for raising funds and managing and trading the attendant risks, providing opportunities for investors, fund managers, and speculators. (Legendary investor E.O. Thorpe once remarked that the difference between gambling and investment is that the latter sounds so much more dignified.) In this section the main mechanisms are described in outline; this is by no means an exhaustive list.

Equities. Issuing stock (or shares) is one of the main mechanisms by which firms raise money for future investment. For a start-up, an IPO (initial public offering) document is drawn up and investors are invited to subscribe share capital at a price set shortly before the IPO goes live. Having committed themselves the investors receive an allocation of shares, and the total amount subscribed is the share capital or equity of the firm. The firm is owned by the shareholders, who expect to receive dividends over time and to participate in the rising value of the shares as the business is built. Post-IPO, the shares are listed with a stock exchange, where they can be traded, invariably on some electronic trading platform; one trading mechanism, the *limit order book* (LOB), is described in Chapter 8. This secondary market is unconnected with the firm itself in that money just passes hands from one investor to another, but it does determine the market value of the firm (equal to the number of shares times share price) and the views of the shareholders are relevant to strategic decisions such as take-over bids. The essence of equity investment is that there are no guarantees beyond the 'limited liability' rules under which the value of a share can never be negative. Purchasers of shares are betting on the success of the firm and, as in most betting, could lose their entire investment (but not more).

Bonds. Another way companies can raise money is by issuing bonds. Purchasers of bonds are lending money (the 'face value') to the company for a fixed period of time, say five years. The company pays 'coupons', generally annually or semi-annually (every six months) which are interest rate payments at an agreed rate, either at a 'fixed' rate, meaning that the rate is fixed at, say, 5 per cent per annum for the lifetime of the bond, or 'floating' if the rate follows current market rates in an agreed way. At maturity the company makes the final interest payment and repays the face value. Once issued, a bond is a tradable asset that can be bought and sold on trading platforms at current market prices. The mechanics of bond valuation are discussed in Chapters 4 and 5. In issuing bonds the company is committing itself to making the scheduled payments on time and in full, regardless of its commercial success. If it fails to do so its credit rating will suffer, and raising further funds will be more difficult and more expensive. The purchaser receives a predictable income stream while he holds the bond but may be able to profit by selling it at an opportune time when market rates are favourable. If the company defaults, the outcome for the purchaser is highly uncertain—he may receive something not far short of the face value or, in the worst case, nothing at all. In short, the 'risk profile' of a bond is very different from that of stocks.

Credit risk. In view of the uncertainties surrounding corporate defaults it seems natural to expect that holders of bonds might wish to insure against this eventuality. Technically, *default risk* is the risk that a bond issuer will not carry out the terms of the bond, that is, fail to make an interest payment, or repayment of the principal amount, on time and in full. Since the mid-1990s, investors have been able to obtain insurance through the mechanism of *credit default swaps* (CDS). In a CDS contract on the default risk of XYZ plc, the investor pays some third party, say ABC, a regular premium over the life of a bond issued by XYZ, based on the perceived credit-worthiness of XYZ. ABC will pay nothing if the terms of the bond are fully carried out, but otherwise pays the investor the difference between the face value of the bond and its post-default recovery value. In this way, holding the XYZ bond plus a CDS contract is almost equivalent to holding a bond with no default risk. It sounds simple, but it is not. To start with, it all hinges on ABC being a much better risk than XYZ. Then there is the size of the market: from modest beginnings it mushroomed to a maximum size (the total face value specified in all active CDS contracts) of $62tn in 2007 but fell out of favour after the financial crisis and is now around a tenth of that size (still a sizable number!). Like the bonds themselves, CDS contracts are tradable assets—this is the difference between them and ordinary insurance—and they violate the cardinal principle of insurance that one should not be able to insure something one does not own. If I can insure my neighbour's house, I have every incentive to burn it down. If other people have insured the same house, they will claim too. Something of the latter sort actually happened in 2005 when the US car parts manufacturer Delphi defaulted, resulting in claims to CDS writers exceeding ten times the face value of Delphi's bonds.

Foreign exchange. The foreign exchange ('FX', 'Forex') market is the largest and most liquid market in the world. Unlike the stock and futures market that are housed in central physical exchanges, the FX market is an OTC ('over-the-counter') decentralized market housed on electronic trading platforms accessible worldwide. The US dollar is involved in 85 per cent of FX trading volume. The Euro is second with 40 per cent, followed by Japanese yen with around 20 per cent. With volume concentrated mainly in these three currencies, FX traders can focus their attention on just a handful of major pairs. The volume of trading is truly extraordinary. In 2016 the average *daily* volume was about \$5tn, roughly 90 per cent of it being generated by currency speculators capitalizing on intra-day price movements. By way of comparison, the daily volume in futures markets was \$450bn, equities \$200bn, the NY Stock Exchange \$30bn. The annual GDP of the USA in 2016 was about \$18.6tn.

Forwards, futures, and options. Now we come to derivative contracts, or just 'derivatives', which are ubiquitous in finance and the focus of much of the academic and practitioner research into correct pricing. 'Derivative' means that the value of the contract depends on, or is 'derived from', the value of one or more *underlying assets* such as stock prices or FX rates.

The simplest such contract is a 'forward'—an agreement to exchange a unit of asset at a specified future time for a price agreed now. Suppose that I am committed to delivering 1,000 shares in Rolls-Royce plc to some counterparty, six months from now. The share price is £9.13 now. I could just wait and purchase the shares on the day, but of course I do not know what price I will have to pay. Another strategy is this. I borrow £9,130 from the bank at an agreed six-month rate of 2 per cent, and buy the shares now. When the delivery time comes, I hand over the shares and clear the loan at the bank by paying £9, 130 × 1.02 = £9, 312.60, which is the price I agree with my counterparty. Effectively, I have fixed a *forward price* of £9.31 for the shares. Could some more subtle strategy result in a lower effective price? We will see in Chapter 3 that the answer to that question is 'no'. In this calculation any dividend payments have been ignored. Since I own the shares, I will receive such payments, so I don't need to borrow quite as much from the bank, reducing the forward price by an amount which, over such a short period, can be accurately estimated. The main point here is that the forward price depends only on the current price, the interest rate, and the dividend. We don't need any kind of model describing the movement of prices to calculate it. It is a 'model-free' result in current jargon.

A common application for forward pricing is to fix FX rates. If the item to be delivered in six months is foreign currency, say €10m for a UK-based company, then an extra ingredient comes into play, namely the Euro interest rate. Instead of purchasing €10m now, I purchase an amount that will accrue to €10m in six month's time when deposited in a Euroland bank. The foreign interest rate plays the same role in the FX case that the 'dividend yield' plays in domestic transactions. The dividend yield is a conventional way to represent dividend payments, in which they are assumed to be paid continuously over time at a rate equal to a fixed fraction q of the current price.

There is no organized market in forward transactions; any existing ones will be bilateral agreements between the parties, and there is an obvious problem: default risk. To continue the Rolls-Royce example, I have agreed to buy shares at £9.31. But suppose that when the moment arrives, the spot price is £9.20—then my best interests would be served by tearing up the forward agreement and buying the shares on the open market. The forward contract, however, specifies that I have the right *and the obligation* to buy at the higher price. A mechanism that preserves the payoff structure of a forward while eliminating the 'walk away' risk is the futures market. In contrast to forwards, futures are exchange traded: the exchange is the counterparty in all transactions and prices are determined by supply and demand. The mechanism is as follows. All participants must open so-called margin accounts at the exchange, which must contain a certain minimum balance. A futures contract at the exchange specifies the asset that it relates to and what the final maturity date T is. Note that the asset need not itself be directly traded. Indeed, some of the biggest futures markets relate to stock indices such as the

S&P500 or FTSE100 indices. While one could in theory trade the index by buying all its component stocks in the right proportions this is totally impractical; trading in the futures contract is an indirect but effective way of trading the index. To simplify exposition, let us just consider once-a-day trading with trading days $0, 1, \ldots, N$ where 0 is the start time of the contract and N corresponds to the maturity date T. Futures prices f_k are published on each day k, and S_k will denote the spot price of the underlying asset. An investor is said to be *long* an asset if the value of her investment portfolio goes up when the price of the asset increases, and *short* if her value goes up when the asset price goes down. The rules of the futures market are

1. To get into the contract on day k_0, the investor simply announces that she is joining and states the number n of units of the contract she wishes to invest in and which side of the market, long or short, she wishes to take. No payments are made by her or the exchange at this point.
2. On each day after k_0, say day j, the investor's margin account is credited with a payment $n(f_j - f_{j-1})$ if on the long side or $n(f_{j-1} - f_j)$ if on the short side. There will be a 'margin call' if the balance ever falls below the minimum allowed level.
3. On the final maturity day N the futures price f_N is *by definition* equal to the underlying asset spot price S_N.

Ignoring the fact that the margin payments are made at different times we see that the total payment per contract, if on the long side, is

$$(f_{k_0+1} - f_{k_0}) + (f_{k_0+2} - f_{k_0+1}) + \cdots + (S_N - f_{N-1}) = S_N - f_{k_0}.$$

(All the other terms cancel out.) This is equivalent to buying one unit of the underlying at price f_{k_0}, the futures price at the time the contract was entered. On the short side the net payment would be $f_{k_0} - S_N$, equivalent to a sale at the same price. The ability to go long or short at the flick of a switch is a key property of a futures market. If one wishes to get out of a futures contract before maturity, one simply enters another contract on the opposite side of the market.

Valuation of futures prices is, in contrast to forwards, not model-free and requires some sophistication. It turns out that the difference between the fair futures price f_t and the corresponding forward price $F(t, T)$ is negligibly small except for futures on very long-term interest rates. In all other cases, forwards and futures are essentially the same thing from an economic point of view.

Finally, options. These deal with the flaw in forward contract mentioned earlier, that the investor has the obligation to follow through with the contract even when it is disadvantageous to do so. An option is a contract between two parties, the holder (buyer) and the writer (seller). Take the simplest example: a call option on a stock with price S_t. This is specified by an *exercise time* T and a *strike price*, or just *strike*, K. The holder pays a fee, the option *premium*, to the writer at inception of the contract and then has the right *but not the obligation* to purchase the stock at time T at price K. The writer, having accepted the premium, has the obligation to deliver one unit of the stock in exchange for a payment K, should the holder wish to make this exchange. The option will be exercised if $S_T > K$, because then the holder is buying the stock for less than it is worth; in principle he could immediately sell it in the market, making a profit of $S_T - K$. (Often, options are 'cash settled': on exercise the writer pays the holder $S_T - K$ in cash.) On the other hand, if $S_T < K$, the holder will not exercise and the option expires worthless. A *put option* is the right to *sell* at price K; it will be exercised when $S_T < K$ with profit $K - S_T$. We write the call and put exercise values as $C(S_T) = [S_T - K]^+$ and $P(S_T) = [K - S_T]^+$ respectively, where $[X]^+$ denotes the greater of X and 0; these are the 'hockey stick' shapes shown in Figure 1.

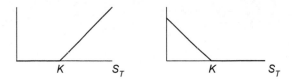

1. **Call (left) and put (right) exercise values.**

It is far from obvious what constitutes a 'fair premium' for a call or put option, a question that was not settled until the decisive results of Black and Scholes (BS) in 1973, which lead to an explosion of activity both in mathematical research and market trading. Once seen as little more than a curiosity, options of various sorts are now an integral part of most financial markets.

Chapter 2
Quantifying risk

As we have seen in Chapter 1, a large part of the purpose of the financial markets is to manage, and indeed profit from, risk—the uncertainty surrounding future events, any event, in fact, that affects the views of investors as to the value of securities.

Figure 2 shows a typical financial price series, in this case weekly values of the FTSE100 stock index over a twenty-year period including the dot-com bubble of 2000–2 and the crash of 2008. The upper chart shows the index levels (say $I_1, I_2, \ldots$) and the lower chart the corresponding returns $R_2, R_3, \ldots$ where $R_k = (I_k - I_{k-1})/I_{k-1}$, the fractional change in the index level. The *market capitalization* of a firm is the total value of its shares (number of shares issued times share price). The FTSE100 index is the total market capitalization of the biggest hundred firms traded on the London Stock Exchange, scaled so that its value when the index was initiated in January 1984 was 1,000.

All financial data looks qualitatively the same, at any time scale (we could have plotted daily or hourly data): erratic and apparently non-stationary behaviour, with 'bursty' volatility (relatively quiescent periods separating periods of more violent moves). A traditional explanation is that price changes are caused by news, that is, occurrence of external events. However, this explanation does not survive analysis: there just isn't enough news to warrant

(a) Index values

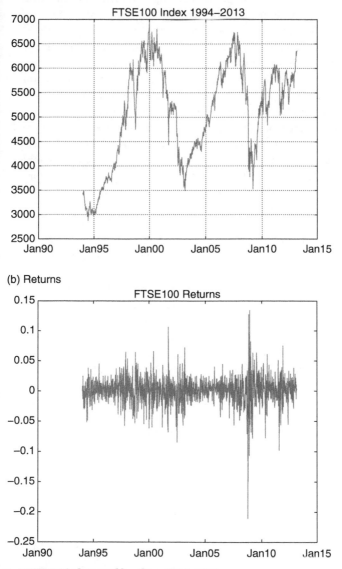

2. FTSE100 index: weekly values, 1994–2013.

price changes at the frequency we observe. So it must be the case that the 'volatility' of prices is in large part internally generated, that is, due to the trading activities of market participants themselves. This is not surprising since market prices are the result of trading, and traders will react to the actions of other traders. It is not to be expected that all market participants will react to news in the same way, and each trader is acting according to her own instincts and agenda. Increasingly, the 'traders' are actually algorithms which, if not well designed, may cause perverse effects including the recently observed 'flash crashes'. In Chapter 3, we will see that there are more fundamental reasons why prices have to be erratic in a well-functioning market and in Chapter 8 we describe the LOB, one of the main trading mechanisms for generating prices. In Chapter 3, specific mathematical models for stock prices will be introduced. Invariably, these are stochastic models, that is, based on probability theory which, since at latest the 17th century, has been the standard framework for dealing with uncertainty. In this chapter we review the main ideas.

The simplest setting, and the one that got probability theory started, is analysis of games of chance. Take die throwing. No-one will contest the idea that, on symmetry grounds (and ignoring manufacturing defects) the chance ('probability') of getting a six or any other specific score on a single throw is 1/6, or that the probability of getting an even number is 1/2. Equally, it is agreed that on two throws the chance of getting a double-six is 1/36: the second throw is unaffected by the result of the first so there are thirty-six equally likely outcomes from throwing the die twice, only one of which results in a double-six. The basic rules of probability follow from this example. We list all the 'elementary events' (in this case, the thirty-six possible ordered pair scores) and assign a probability to each elementary event. The probabilities are numbers between 0 and 1 that add up to 1. 'Events' are just collections of elementary events, and the probability of an event is the sum of the probabilities of the constituent elementary events. In our example, 'the sum of the two scores is at least 9' is an event consisting of ten elementary events, so its probability is 10/36. It is clear that if we have two mutually exclusive events (i.e. they contain no common elementary event) then the probability that one or other occur is the sum of the individual probabilities. A further concept is *independence*. We say that two events are independent if the probability that both occur is the product of the individual probabilities. That is how we got 1/36 as the probability of double-six. In summary, we have the *basic rules of probability*:

(i) Each elementary event is assigned a probability, a number between 0 and 1.

(ii) The probability of any event is the sum of the probabilities of the elementary events comprising the event.

(iii) Something must happen: the sum of the probabilities of all elementary events is 1.

A *random variable* is a rule that associates a number with each elementary event. We could, for example, envisage a game in which a die is caste twice and a player loses £1 if the sum of the scores is less than nine and wins £2.60 otherwise. The corresponding random variable takes the value 2.6 on the winning set {36, 45, 46, 54, 55, 56, 63, 64, 65, 66} and -1 on the remaining elementary events, of which there are twenty-six. The *expected value* (or *mean* or *expectation*) of a random variable is obtained by summing the value times the probability over all the elementary events. In the example the probability is 1/36 for every event, so the expected value is $2.6 \times 10/36 - 1 \times 26/36 = 0$. This is a *fair game*, encapsulating the idea that betting on an unlikely event should yield a large reward if that event occurs.

Now for 'continuous' random variables, that is, analysis of quantities such as weight, height, or stock exchange prices which take values in a continuous range of numbers. Actually, stock exchange prices are technically speaking not continuous, because the values are always quantized to some minimum price move, known as a 'tick'; for the S&P500 stock index in New York the tick size is 1 cent, so all prices are integer numbers of cents, but for most purposes it is far too cumbersome to consider them as such, and much more convenient to analyse them as continuous variables. The most convenient way to represent probabilities for continuous variables is through *density functions*. For example, the formula for the density function of the *normal distribution* $N(m, \sigma)$ is

$$f_{m,\sigma}(x) = \frac{1}{\sigma\sqrt{2\pi}} \exp\left(\frac{(x-m)^2}{2\sigma^2}\right). \tag{1}$$

It has two parameters, the *mean m* and the *standard deviation σ* (the Greek letter sigma, by long-standing convention). The mean, or *expectation*, is the 'centre of gravity' of the density function and the standard deviation measures how widely around the mean the density is dispersed. Figure 3 shows two normal density functions, both having mean $m = 0$ while the standard deviations are $\sigma = 1, 2$.

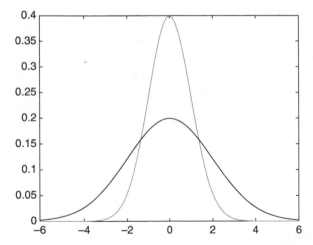

3. **Normal density function with variance 1 (narrow shape) and 2 (wide shape).**

If, in general, X is the value of some random quantity with density function f then for any numbers $a < b$ the probability that X lies in the interval between a and b is by definition the area underneath f between those two values. In particular the total area under the density function must be 1. The probability that X is exactly equal to a specific number c is zero, because this probability cannot be more than the probability that X lies in the interval between a and b for any a, b such that $a < c < b$, and the area in question shrinks to zero as the distance between a and b is reduced. This means that we cannot compute expectations by summing values times probabilities as before. The right formula for the expectation of a random variable with density function $f(x)$ is

$$\mathbb{E}[X] = \int_{-\infty}^{\infty} x f(x) dx.$$

Not every random variable has a density function but each has a *distribution function* F whose value at a point a is the probability that the corresponding random variable X is less than or equal to a. If X has a density function then $F(a)$ is just the area under the density function to the left of a. Figure 4 shows the distribution function for a single-die throw, where X is the score obtained.

The standard deviation for a general random variable X with mean $\mathbb{E}[X] = m$ is defined as follows: let $Y = (X-m)^2$, the square of the distance from X to m. Then $\sigma^2 = \mathbb{E}[Y]$, the 'mean squared deviation from the mean'. As in the case of the mean, not every random variable has a standard deviation. We see from (1) that a

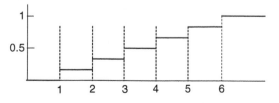

4. **Distribution function for one-die throw.**

normal distribution is completely specified by giving its mean m and standard deviation σ.

Given two random variables X_1 and X_2 with means m_1, m_2 and standard deviations σ_1, σ_2, the *correlation* between X_1 and X_2 is $r = \mathbb{E}[(X_1 - m_1)(X_2 - m_2)]/(\sigma_1\sigma_2)$. This is a number between -1 and $+1$; if $r > 0$, X_1 and X_2 are generally moving in the same direction while if $r < 0$ they tend to move in opposite directions. They are 'uncorrelated' if $r = 0$. Independent random variables are always uncorrelated, but the converse is not generally true. Uncorrelated normal random variables, however, are independent.

Probability theory may seem esoteric, but it has a clear connection with observable reality through two of its most basic results, the *strong law of large numbers* and the *Central Limit Theorem* (CLT). These were uncovered in the 18th century by de Moivre and Laplace but, perhaps surprisingly, the final versions were not arrived at until the 20th century, following development of the now universally accepted mathematical theory of probability by A.N. Kolmogorov in 1933. Both concern the behaviour of i.i.d. (independent, identically distributed) random variables such as the scores in successive throws of a die. Denoting these random variables $X_1, X_2, \ldots$ the assumptions are that they all have the same distribution, with common mean m, and are independent. Let Y_n denote the average of the first n variables: $Y_n = (X_1 + X_2 + \ldots X_n)/n$. Then Kolmogorov's strong law of large numbers states that with probability 1 the sequence Y_n converges (i.e. gets closer and closer) to the mean m. Note that each Y_n is a random variable but m is a fixed constant. Figure 5 shows the behaviour of Y_n when the X_k's correspond to die throws; here $m = 3.5$.

For the CLT we need additionally to assume that the random variables X_k have a (common) standard deviation σ. Define $Z_n = \sqrt{n}(Y_n - m)$, which amounts to 'centring' the X_k by subtracting the mean and then dividing the sum by $\sqrt{n}$ rather than n. In this case the Z_n sequence does not converge to a

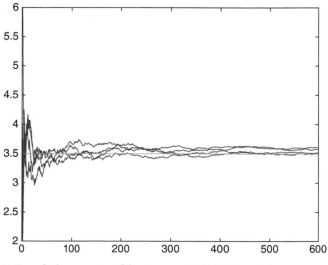

5. Cumulative averages of the scores on 600-die throws (four samples).

number, but 'converges in distribution' to the normal distribution with mean 0 and standard deviation σ. This means that, given any number a, for large n the probability that Z_n is less than a is close to the value predicted by the normal distribution with mean 0 and standard deviation σ. Figure 6 demonstrates the effect when the random variables are die throws. We simulate 150-die throws and compute Z_n for $n = 150$ as before. Then we repeat the process twenty times, giving twenty independent samples of Z_n, and for a range of numbers x between -3 and 3 we plot the proportion of these samples whose values are below x, giving the dotted curve in Figure 6. Increasing the number of samples to fifty gives the dash-dot curve, and increasing it further to hundred gives the solid curve. The normal distribution function is also shown. As can be seen, the hundred-sample curve is getting pretty close to the normal distribution in the central region, although more data would be needed to capture the left and right ends where only a small proportion of the samples lie.

19

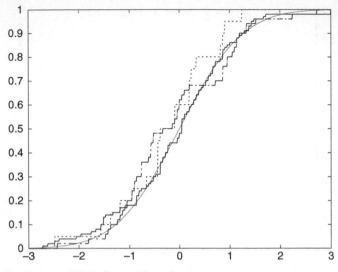

6. The central limit theorem in action.

The *frequentist approach* to probability and statistics relies on the two basic results we have discussed: the law of large numbers and the CLT. It is applicable when the data is in the form of repeated independent trials, for example in analysis of insurance claims for car accidents, or medical trials where randomly selected patients are given treatments or placebos. There are, however, many other situations where the necessary conditions of repeatability and independence are not met. Examples range from weather forecasting (estimating the probability of rain tomorrow) to political questions such as the probability that a particular candidate will win a presidential election, to financial questions such as estimating the probability that the issuer of a bond will default on the payment of coupons or principal. In such cases the interpretation of probability is more controversial. There is an apparently simple way out: the subjectivist point of view that assignment of probability is just an expression of the investigator's opinion and not an objective truth to be shared by everybody. This raises, however, obvious questions. If I predict that the probability of rain tomorrow is 25 per cent, and rain the next day 60 per cent, then I am, at a minimum, asserting that rain tomorrow isn't particularly likely but rain the next day much more likely. But what is the quantitative significance of the actual numbers, and why do I need a sophisticated mathematical theory such as Kolmogorov's to back them up?

An answer to these questions was provided in 1926 by Frank Ramsey, an astonishingly precocious researcher in the circle of Keynes, Russell, and Wittgenstein who tragically died aged 27 in 1930. Similar ideas were independently introduced, and developed in much greater detail, by Bruno De Finetti in the 1930s. Their ideas are extremely close to central topics in mathematical finance developed from the 1970s onwards, although this connection has been largely overlooked in the mathematical finance community. They relate to gambling strategies. To get started, consider the odds displayed in Table 1 for a horse race at the Tipperary track in Ireland in 2016. The odds on Now or Never are 6 to 5, meaning that a €5 bet pays €6 plus return of the stake if Now or Never wins. If the odds are x to y then x is the net payoff on a win, on a stake of y; see Table 2.

Table 1 Betting odds for the 5.40 Tipperary, 25 August 2016

Horse	x	y	Imp prob	Stake	Net win
Now or Never	6	5	0.455	45.45	100
Creggs Pipes	7	4	0.364	36.36	100
Tanaza	11	2	0.154	15.38	100
Lina de Vega	10	1	0.091	9.09	100
World of Good	125	1	0.0079	0.79	100

Table 2 Payments on an x/y bet

	Before race	After race	Net payoff
Win	$-y$	$x + y$	x
Lose	$-y$	0	$-y$

Are the odds quoted for Now or Never 'fair'? Betting on Now or Never is like flipping a coin and winning €6 on heads while losing €5 on tails. This is fair if the expected loss equals the expected gain, which will be the case if the win probability is 5/11 and the loss probability 6/11. The extra euro on the win side is compensated by a win probability slightly less than 1/2. We call 5/11 the 'implied probability', and its value for each of the horses is shown in the fourth column of Table 1 (algebraically, its value is $y/(x + y)$). We say that the odds are *consistent with probabilities* (or just 'consistent') if these implied probability assignments obey the basic rules of probability as enunciated earlier.

At the racetrack, exactly one horse must win and 'Now or Never wins' is an elementary event. But the sum of the implied probabilities in Table 1 is 1.0709, so the odds are *not* consistent. An interesting property of the implied probabilities is that if one places bets proportional to them then the amount paid out on a win will be the same whichever horse wins. This is shown in the last two columns of Table 1, where the stakes are a hundred times the implied probabilities. Placing all these bets guarantees a payout of 100 at the end, but unfortunately the total stake is 107.09, guaranteeing a profit of 7.09 to the bookie. Note, however, that if Lina de Vega were to be taken out of the race and the bookie failed to adjust the odds on the remaining horses then the strategy guaranteeing 100 payout would cost 98. In finance, this is called an *arbitrage opportunity*—the existence of a strategy that wins for sure, with no possibility of loss. A market is *arbitrage-free* if there are no arbitrage opportunities.

Now let us turn to the theories of Ramsey and De Finetti, analysing 'subjective probability' or probability as *degree of belief*. These are based around so-called Dutch Book arguments (DBA—nobody seems sure how this terminology arose) which concern the existence or otherwise of arbitrage strategies. Their reasoning is as follows: if a person ('the bookie') has beliefs about future outcomes, then he should be prepared to accept bets from others at odds consistent with those beliefs. But then we have the *Dutch Book Theorem* (DBT): if the implied probabilities corresponding to these odds do not satisfy the basic rules of probability then the bookie can be 'Dutch booked', meaning that, at these odds, bets can be placed that guarantee a loss to the bookie. Hence the bookie is being irrational in offering these odds; a rational person must adhere to the tenets of probability theory.

Debate rumbles on in philosophical circles as to the merits of DBA; a lengthy discussion can be found in the online *Stanford Encyclopedia of Philosophy*. One technical point is that validity of the theorem depends on the ability to place both positive and negative bets at the same odds, contrary to the practice in any real market. In the Tipperary example there would indeed be a Dutch Book strategy were negative bets allowed: just reverse the sign of the strategy given in Table 1, so that the bookie pays out €107.09 in advance and collects the 'winnings', always equal to €100, after the race. Our purpose here, however, is not to follow up technicalities but merely to point out that the classic DBA, translated into financial terminology, is saying that the bookie is creating an arbitrage opportunity by offering inconsistent odds. The DBT is in fact a special case of the central result in mathematical finance, the so-called Fundamental Theorem of Asset Pricing (FTAP), which is discussed in Chapter 3. A key feature of the result is that *all bets under consistent odds are fair*, that is, the expected payoff of the bet is zero. This is another fact that plays a key role in financial applications.

Table 3 Revised fair odds: Lina de Vega quoted at 49 to 1

Horse	x	y	Implied prob
Now or Never	6	5	0.455
Creggs Pipes	7	4	0.364
Tanaza	11	2	0.154
Lina de Vega	49	1	0.020
World of Good	125	1	0.0079

In our example there are only five elementary events, corresponding to each horse being the winner. In the case of fair odds, it is easy to see how to propose fair odds for a wider range of bets. By way of illustration, let us revise the odds on Lina de Vega to 49 to 1 (see Table 3), so that overall the odds are consistent: the implied probabilities sum to 1. Suppose we want to offer a place bet on Now or Never (i.e. bet that Now or Never comes in first or second). There are many ways to do so. We now have twenty elementary events, as there are four possible horses in second place for each of the five possible winners. If we arbitrarily assume that, for each winner, all four second-placers are equally probable, then the probability that Now or Never places is $0.455 + (1 - 0.455)/4 = 0.591$, corresponding to odds of roughly 4 to 5. The bookie could not be Dutch booked if these odds were offered. This kind of question comes up frequently in finance: we have some contracts priced such that there is no arbitrage. If we wish to introduce another contract, what range of prices can we quote such that the market is still arbitrage-free when the new contract is included?

Stochastic modelling

Stochastic modelling means specifying a set of random variables that accurately represents some data presenting unpredictable features. Look at the FTSE100 Index series in Figure 2(a), for example. As presented, this is a once-a-week sampling of values that are recomputed many times during trading hours, based on prices that change on a second-by-second basis. There is a basic distinction in stochastic modelling between models that represent the evolution of prices continuously over time and those that represent sampled values at some specified frequency ('discrete-time models'). In finance, the former are generally used in option pricing and the latter in econometrics and risk management. We hope, in this chapter and the next, to unveil why this is so.

Let us start with the discrete-time weekly returns $X_0, X_1, \ldots$ of Figure 2(b). The simplest model would be to state that these are i.i.d. random variables, but this does not seem credible in the current instance, since it appears that at the very least the variance is not constant. If we reject the i.i.d. hypothesis then almost invariably the approach to developing a more general model is as shown in black-box form in Figure 7.

The inputs are the initial value X_0 and a sequence of i.i.d. random variables $W_1, W_2, \ldots$ with zero mean. The variable W_k represents the 'surprise' at time k—the part of X_k that cannot be predicted from prior observations. At time 1, W_1 is input and the black-box algorithm—depending on some parameters $b_1, \ldots, b_n$—spits out a value for X_1. At each subsequent time the next W_k is input and the corresponding X_k, which may also depend on previous data stored in memory, is computed. To get the picture, consider a simple

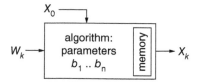

7. **Time series modelling.**

example, the *first-order autoregression* AR(1) with the W_k being i.i.d. $N(0, 1)$ random variables and X_0 is $N(m, \sigma^2)$. The model is

$$X_k = b_1 + b_2 X_{k-1} + b_3 W_k, \quad k = 1, 2, \ldots \quad (2)$$

The coefficient b_3 is the *volatility*. In this equation, $X_1 = b_1 + b_2 X_0 + b_3 W_1$, which is a sum of normal random variables, and it is a property of the normal distribution that such sums are also normal. Similarly, all the X_k are normal, so to get the distribution we only have to compute the mean m_k and variance σ_k. Taking expectations in (2) we see that $m_k = b_1 + b_2 m_{k-1}$, $m_0 = m$, and similarly $\sigma_k^2 = b_2^2 \sigma_{k-1}^2 + b_3^2 \sigma^2$, so m_k and σ_k can be computed recursively. In the 'stable' case $|b_2| < 1$ we see that, for large k, $m_k \approx b_1/(1 - b_2)$ and $\sigma_k \approx b_3 \sigma / \sqrt{1 - b_2^2}$. So for large k the X_k all have the same distribution; the output is 'stationary'. The big thing that the extra structure brings us is *correlation*. In fact the large-k correlation coefficient between X_k and X_{k-1} is just b_2 and we can easily work out the correlation between any two X_j, X_k.

In finance, the purpose of the model is usually either (a) to compute the distribution of X_k, or some statistic of the distribution such as the variance or a quantile (see Chapter 7), or (b) to compute the expectation of some function $Y = f(X_k)$, as in option pricing, Chapter 3. In simple cases such as (2) one can calculate the distribution of X_k mathematically and then compute directly the required statistic or expectation. Most models are, however, too complex for this, and the universal fall-back option is simulation or 'Monte Carlo'. Standard numerical software such as Excel or Matlab will generate independent samples of the normal distribution. Plugging these samples into (2) as values of W_1, $W_2, \ldots$ gives us a random value for X_k. Repeating this process 100,000 times gives us 100,000 independent samples of X_k, enough to give us a good idea what the distribution of X_k is and, by the Law of Large Numbers, what the expectation $\mathbb{E}[Y]$ is. Is (2) a good model for the FTSE return series? No, because the wide variation in the size of the returns suggests that the volatility should not be a constant. To allow for this we can use the very widely applied volatility model known as GARCH(1,1), in which the volatility parameter b_3 in (2) is replaced by the time-varying parameter σ_k satisfying

$$\sigma_k^2 = b_4 + b_5 X_{k-1}^2 + b_6 \sigma_{k-1}^2. \tag{3}$$

Thus the volatility at any time depends on the volatility and the actual return at the previous time. No extra 'noise' is introduced. Further modifications of the model may be made by choosing different distributions for the random inputs W_k. A huge amount of material dealing with methods for estimating the parameters $b_1, \ldots, b_6$, and with 'significance' tests aimed at determining whether or not a proposed model is an acceptable description of the data, can be found in the econometrics literature.

In Chapter 3 we address problems of option pricing, and here the approach to modelling is quite different. As will be seen, option prices are expressed as discounted expectations of the option payoff. The most basic requirement of a model is that it correctly prices market-traded options whose prices—arrived at by 'supply and demand'—are available market data. A model is *calibrated* if it meets this requirement. Calibration is an optimization procedure: if there are M options with prices $p_1, \ldots, p_k$ and $b = (b_1, \ldots, b_n)$ are the parameters of some stochastic model such that the model price for the jth option is $p_j^M(b)$ then calibration is achieved by minimizing over b the total pricing error $\sum_{j=1}^{k} (p_j - p_j^M(b))^2$. If the minimum is not extremely close to zero then the model is inadequate and a rethink is called for. It turns out that a pricing model does not necessarily have to be particularly accurate in econometric terms. For example, the model used by BS would not survive any statistical tests for adequacy in representing data such as the FTSE index in Figure 1, but nonetheless it is a cornerstone of option pricing theory; we explain why in Chapter 3. It is, however, true that different models, all calibrated to the same data, will give different prices for a putative new option.

In the aftermath of the 2008 financial crisis studying the effect of model error became a hot topic of research. Many participants in this endeavour describe the risks posed by relying on inadequate models as *Knightian uncertainty*, referring to ideas of Frank Knight in a famous book entitled *Risk, Uncertainty and Profit* published in 1921 and still in print, even though the first sentence in the book is: 'There is little that is fundamentally new in this book'. But in fact his ideas are far more radical. He set out to solve a puzzle in economics: why is it that in a competitive market prices are not driven down to the cost of production, leaving no profit at all? Knight's answer is in three parts.

(i) If we are in a static economy where there is no risk and no technological progress, the prices will indeed be driven down in this way.

(ii) If there is mild randomness in the form of business risks and random but evolutionary technological progress then entrepreneurs can hedge against these factors, leaving themselves in the same position as Case (i).

(iii) Therefore the only way entrepreneurs can consistently make profits is by taking a series of unquantifiable risks by launching new products. The perfect example is the Apple iPad. It was not just a new product; it was a new *category* of product. There was nothing to compare it with and no significant amount of market research could be carried out without destroying the essential element of surprise.

The same point was taken up in 1937 by John Maynard Keynes:

By 'uncertain' knowledge, let me explain, I do not mean merely to distinguish what is known for certain from what is only probable. The game of roulette is not subject, in this sense, to uncertainty; nor is the prospect of a Victory bond being drawn. Or, again, the expectation of life is only slightly uncertain. Even the weather is only moderately uncertain. The sense in which I am using the term is that in which the prospect of a European war is uncertain, or the price of copper and the rate of interest twenty years hence, or the obsolescence of a new invention, or the position of private wealth-owners in the social system in 1970. About these matters there is no scientific basis on which to form any calculable probability whatever. We simply do not know.

It could be argued that the failure to distinguish between risk and uncertainty was a factor in the financial crisis. As we discuss in Chapter 5, a major problem in credit risk is lack of data, and it seems clear in retrospect that models were built, and prices offered, in situations where there was 'no scientific basis'. At the same time, it is also true that companies offering long-term contracts of life insurance and pensions have survived for centuries. The secret? Adequate reserves and hedging—see Chapter 6.

Chapter 3
The classical theory of option pricing

The 'classical period' of mathematical finance began on 29 March 1900 when Louis Bachelier defended his PhD thesis and ended on 15 September 2008 when the Wall Street bank Lehman Brothers filed for Chapter 11 bankruptcy. During this period a more-or-less complete theory of *arbitrage pricing* was developed, explanation of which is the subject of this chapter. The underlying idea is closely related to the DBA described in Chapter 2 but brings in a new factor: prices in financial markets evolve over time and participants are able to trade at any time, instead of just taking bets and awaiting the result as at the Tipperary racetrack. In addition to the general theory, pricing models and methods have been developed for specific markets—FX, interest rates, credit—and some of these are discussed in subsequent chapters.

Louis Bachelier's thesis was entitled 'Théorie de la Speculation' and his inquisitors were a jury of eminent mathematicians at the Sorbonne in Paris. The topic of the thesis was the valuation of financial options, which to the jury was, to say the least, an extraordinary choice. They passed him with the grade of *honoré*, apparently the highest that could be awarded for a thesis that was deemed to be essentially outside mathematics and to include arguments that were not rigorously justified. They under-estimated Bachelier: his thesis was ahead of its time, and he introduced several constructs in probability that are now known by the names of better known scientists who independently rediscovered them later.

Call options were defined in Chapter 1. The purchaser pays a premium now for the right, but not the obligation, to purchase at some later time T one unit of a certain traded asset, the 'underlying asset' S_t, at the strike price K. The option holder will exercise only if $S_T > K$, so the exercise value is $[S_T - K]^+$, the 'hockey stick' payoff function of Figure 1.

Bachelier was the first person to come up with an evaluation of the premium. His analysis was based on the idea of market balance: in any transaction on the market, there has to be a buyer for every seller, so market prices cannot consistently favour one over the other. Bachelier summarized this in an impressive-sounding dictum

L'espérance mathématique du speculateur est nulle

('The speculator's expected gain is zero'). This is exactly the same as the concept of 'fair odds' or 'consistent pricing' that we were examining in Chapter 2. Since the underlying asset is itself traded, its expected price at some later time must coincide with today's price. In the case of an option, the purchaser has already paid the premium and the payoff on exercise is always either zero or some positive amount, so the fair premium must balance the expected value of that payoff, so that the holder has *espérance nulle*. The remaining question is: under what probability distribution should we take the expected value? So far we just know that its expected value is equal to today's price. Bachelier introduces assumptions that amount to saying that the market 'looks the same' all the time: price moves over non-overlapping time intervals, say this week and next week, have the same distribution, and are independent. He then concludes that the price moves must have the normal distribution with mean 0 and variance proportional to the length of the interval. This leaves only one parameter to be determined, the constant of proportionality, or volatility, which can be estimated from historical price data. Then the option value can be calculated simply by integrating the option payoff function of Figure 1 against the normal density, and this is Bachelier's option pricing formula. Given Bachelier's price model, his formula is correct—but not for the reason Bachelier thought it was. Note that Bachelier's is a continuous-time model: his arguments depend on considering price moves over arbitrarily short time intervals.

In 1905, Albert Einstein came up with exactly the same model as Bachelier, in a different context but using mathematical arguments that closely paralleled Bachelier's. Einstein was analysing the irregular movements of microscopic particles suspended in a liquid, originally observed in 1827 by Scottish botanist Robert Brown. Einstein was looking for direct evidence for the molecular theory of heat. He showed that random motions similar to those observed by Brown were consistent with this theory, and that the random displacements over non-overlapping time intervals would be independent. He also concluded, just as Bachelier had, that the mean square displacement of a particle in an interval of time is proportional to the length of the interval. However, in the physics context this conclusion has an unexpected consequence: if the mean *squared* displacement is proportional to time, then the mean *absolute* displacement is proportional to the *square root* of the time; see Figure 8 where the curve is the square root function. The average *speed* of the particle is distance/time. Points A and B on the curve correspond to times 0.25 and 0.56 and the corresponding average speeds are the slopes of the lines, also shown, joining A and B to the origin. As can be seen, the average speed increases as the time is reduced, and there is no limit: for any given speed—such as the speed of light—there is some time below which the average speed exceeds that number.

Einstein explains that his analysis is based on assumptions that cannot be valid below some minimum time or space threshold.

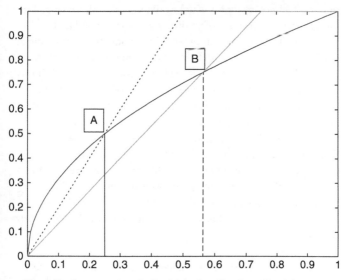

8. Mean displacement (vertical axis) vs time (horizontal axis).

That is satisfactory for him, but for the finance application, where physical constraints are irrelevant, it is important to know whether it is possible to specify a mathematical model for prices in which the prices increments over arbitrarily short time intervals are normally distributed and exactly independent. A positive answer to this question was given was given by Norbert Wiener in 1923; the term 'Brownian motion', also known as the 'Wiener process', now usually refers to this mathematical construct rather than the phenomenon observed by Brown. Wiener's result is a cornerstone of modern finance, as we shall see.

Fellow mathematicians were aware, and appreciative, of the mathematical innovations contained in Bachelier's thesis and subsequent work, but the economic side of his work languished, totally unrecognized. Options markets were small and specialized, his approach would have been too abstract to be appreciated by traders, and running an options market on any scientific basis requires technology that was not available. All of this changed in the 1960s when the scale of trading started to expand and economists became more interested in the operations of financial markets. This is where Paul Samuelson comes into the picture. The catalyst for igniting his interest in Bachelier was L.J. ('Jimmy') Savage, a leading mathematical statistician of the day. Savage had read Bachelier, and sent postcards to his economist friends, including Samuelson, telling them they should do so too. Samuelson had been engaged in what he later described as 'desultory researches' into option pricing, and Bachelier's work gave him a new perspective. Robert Merton, later to receive the Economics Nobel Prize jointly with Myron Scholes for their work on option pricing, joined the MIT team in 1967. Their co-conspirator Fischer Black had sadly died before the Nobel prize was awarded.

The decisive breakthrough, strikingly original though building on the earlier investigations, was the publication of the BS formula in 1973. The authors announce the key idea right away in the abstract to their paper:

> If options are correctly priced in the market, it should not be possible to make sure profits by creating portfolios of long and short positions in options and their underlying stocks.

The reader will recognize the 'DBA' in this statement, but also the extra ingredient of trading. A basic assumption is that the market is 'frictionless', meaning that assets can be bought or sold at the same price in arbitrary amounts. Of course, no real market is frictionless, but making this assumption substantially simplifies the analysis, and the small adjustments occasioned by transaction costs can be estimated later. With these assumptions and with the price model proposed by Samuelson, BS were able to show that, for a given call option, a trading strategy can be created that perfectly replicates its exercise value (Figure 1) whatever the price moves of the underlying asset. Then the initial investment required to set up this portfolio must be the price of the option, otherwise we would have two different prices for exactly the same thing, creating a 'sure profit'.

Binomial models

Rather than studying the continuous-time BS model, closely related to Bachelier's, right away, it is better first to understand the ideas in the simplest possible setting, the so-called binomial model introduced by Cox, Ross, and Rubinstein in 1979. The starting point is the one-period model shown in Figure 9. Although at first sight absurdly artificial, this model has the big advantage that the whole theory can be described in a couple of pages and the only calculation required is the solution of a pair of simultaneous equations.

The model is shown in Figure 9. At time 0, an asset ('stock') has price s. At time 1, its price either moves up by a factor $u > 1$ or down by a factor $d < 1$. The other asset in the market is a bank account paying interest at per-period rate r, so that £1 deposited at time 0 accrues to £$R = 1 + r$ at time 1. An option is written on the stock, exercised at time 1 with values a, b in the up and down states respectively, see Figure 9. For example, if the option were a call option with strike K between us and ds, then a would be equal to $us - K$ while b is zero. There is arbitrage in this model if $R \leq d$ or $R \geq u$: in these cases borrowing from the bank and investing in stock, or *vice versa*, realizes a riskless profit. We therefore suppose that $d < R < u$.

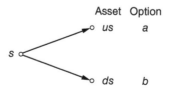

9. **Single-period binomial tree.**

At time 0 we form a portfolio consisting of £B in the bank and N shares of stock. The value of this portfolio is £$(B + Ns)$ and its value at time 1 will be either $RB + Nus$ or $RB + Nds$. Now choose B, N to satisfy

$$RB + Nus = a \tag{4}$$
$$RB + Nds = b,$$

to which the solution is

$$N = \frac{1}{s}\frac{a - b}{u - d}, \qquad B = \frac{bu - ad}{R(u - d)}. \tag{5}$$

With this choice of N, B the value of our portfolio coincides with the option exercise value, whichever way the price moves. We say the portfolio *replicates* or *perfectly hedges* the option payoff. Hence the value of the option at time 0 must be equal to the value V of the portfolio at time 0, which is

$$V = B + Ns = \frac{1}{R}\left(\frac{bu - ad}{u - d}\right) + \frac{a - b}{u - d}. \tag{6}$$

The argument is an application of the portentously named Law of One Price, which says that, in a frictionless market, two contracts that deliver exactly the same cash flows at all points in the future must have the same value now—otherwise one could sell one of them, buy the other, and walk away with an arbitrage profit.

We have shown that there is a unique arbitrage-free price for the option, obtained by calculating the 'perfect hedging' strategy (B, N). This is the essence of the BS argument. However, more can be said. The reader should check that the price formula (6) can re-arranged to read

$$V = \frac{1}{R}(qa + (1 - q)b), \tag{7}$$

where $q = (R - d)/(u - d)$. Note that q does not depend on the option contract and that our no-arbitrage assumption $d < R < u$ is equivalent to the statement that $0 < q < 1$. We can therefore interpret q, $(1 - q)$ as probabilities of an upward or downward move respectively, and rewrite (6) again as

$$V = \frac{1}{R}\mathbb{E}_q[G],$$ (8)

where G is the 'random variable' taking values a, b on an up or down move, and $\mathbb{E}_q$ denotes expectation. Otherwise stated: any option price is the expected value, under the same probability q, of its discounted payoff. The 'discount factor' $1/R$ arises because the option premium is paid at time 0, while the payoff occurs at time 1. If (as in Bachelier's day) the premium were paid at the end, the discount factor would disappear; then the speculator's expected gain is indeed zero, since the expected gain from payoff is exactly matched by the premium paid. The probability determined by q is called the *risk-neutral* probability; this terminology is explained in Chapter 6. For a further characterization of q, consider setting $a = us$ and $b = ds$, so the 'option' payoff G is actually just one unit of the stock (write $G = S$). From (5) we see that $N = 1, B = 0$, so that $V = s$ or, from (8),

$$s = \frac{1}{R}\mathbb{E}_q[S].$$ (9)

This shows that q is the unique probability such that the market is a 'fair game': today's price is the expected discounted value of tomorrow's price, the discount factor $1/R$ arising because of the different payment times.

The key thing to realize here is that *no probability was specified in advance*: all we stated was that tomorrow's price must take one of two possible values u, d. There is no claim that q is the actual probability of an upward move; what we have shown is that options are correctly priced by the formula (6), and that is because this formula gives the set-up cost of a perfectly replicating portfolio. A further conclusion is that this is a 'complete market', meaning that there is only one possible choice for q and any option has a unique arbitrage-free price.

Aside from the question of discounting, which stems from the fact that, when interest rates are not zero, a pound today is not the same thing as a pound tomorrow, the above results are just an application of the DBT; we have two 'bets'—the underlying asset and the option—and the market can be Dutch booked unless the odds (read: prices) are consistent with a probability model. The new part is the perfect hedging idea, and this will come into its own when we consider multi-period models. To summarize, we have shown:

1. The model is arbitrage-free if and only there is a q strictly between 0 and 1 such that the market is a fair game in the sense that today's price is the discounted expectation of tomorrow's price.
2. When the model is arbitrage-free, any option has a unique value.
3. This value is the initial investment required to replicate the option by trading in the market.
4. The value can be expressed as the discounted expectation of the exercise value, calculated under q.

The next step is to extend the argument to multi-period models, but before doing so let us briefly stop to consider the single-period trinomial model shown in Figure 10, which provides an example of an *incomplete* market.

The asset price now takes one of three values at time 1. To construct a hedging portfolio for an option in the binomial case we solved equations (4), in which there are two equations and two unknowns B, N. Here we have three equations but, if we are trading in the underlying asset and cash, still the same two

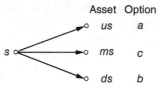

10. Single-period trinomial tree.

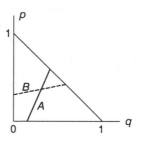

11. Calibration of trinomial model.

unknowns. In general there is no solution and perfect replication is impossible. What about arbitrage? To specify a risk-neutral distribution now requires two numbers q and p so that the up, middle and down moves have probability q, $(1 - q - p)$ and p respectively, with $q + p \leq 1$ to ensure that the middle term is positive. Risk-neutrality means that the 'fair odds' condition (9), here $s = \mathbb{E}_{p,q}[S]/R$, is satisfied and this restricts p, q to lying on a line such as A in Figure 11, determined by the model parameters. The corresponding option value is $V = \mathbb{E}_{q,p}[G]/R$, and this is an arbitrage-free price since consistent odds have been constructed. It is not, however, the only arbitrage-free price since different combinations of q and p are possible; we say the market is *incomplete*. If somebody were to offer the option at a specified price V_1 we could go through the above process again to determine what values of q, p are consistent with this price, giving another line, say B in Figure 11. If, as in the figure, A and B intersect, then the point of intersection is the unique q, p that consistently price both the underlying and the option. The market is complete if we are able to trade in all three assets, the bank account, the underlying and the option (with price V_1). On the other hand, if A and B fail to intersect then offering the option at V_1 creates an arbitrage. In general: if there are n possible price moves then we are going to need n traded assets plus the bank account to 'span the market'.

Finally, we move to Cox–Ross–Rubinstein multi-period binomial setting where trading may take place over several time periods before options are exercised. This model was of key importance in turning the BS result from a theoretical construct to something that could easily be implemented in a few lines of computer code. It also provided an answer to the question of how to value American options, to which we return later in this chapter. The three-period binomial tree is illustrated in Figure 12 and represents the evolution of the underlying asset price, which can move up or down in each time period. The initial price is s, and in the first period it moves up to $us > 1$ or down to ds where $u > 1$ and we take $d = 1/u$. In subsequent periods the current price is multiplied by these same factors. Since $u \times d = 1$, an up move followed by a down move brings us back to where we started, as the figure shows. For example, if $u = 1.1$, corresponding to a 10 per cent price rise, then $d = 0.909$, a fall of 9.09 per cent. One consequence of this structure is that, as in real life, but unlike Bachelier's model, the price can never be negative. As before there is a bank account paying interest at rate r per period, and $R = 1 + r$ satisfies $d < R < u$ to avoid arbitrage. Looking at just the first

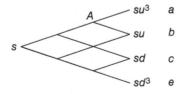

12. Three-stage binomial tree.

period, the risk-neutral 'up' probability q is equal to $(R - d)/(u - d)$ as we calculated earlier, equal here to $(Ru - 1)/(u^2 - 1)$, and an option exercised at time 1 paying a on an up move and b on a down move has value $(qa + (1 - q)b)/R$ at time 0. But now suppose we are at node A on the tree, that is, at time 2 having had two upwards moves. Then the situation that faces us for the remaining period is just the same as at time 0—the next price will be the current price multiplied by either u or d—except that the current price is now su^2, not s. The risk-neutral probability starting at A is the same number q as at time 0, and the same is true for any 'triangle' in the tree. Consequently the value needed at node A to replicate an option paying amounts a, b, c, e at time 3 is $V_A = (qa + (1 - q)b)/R$, with similar expressions for V_B, V_C, the values at the other two time-2 nodes. But now we have reduced the problem to replicating an 'option' paying V_A, V_B, V_C at time 2. By the same argument we can reduce the problem again to replication of known amounts at time 1, which is just the one-period case to which we know the answer.

Under risk-neutrality, every up/down move has probability q, $(1 - q)$. We can generate random 'price paths' such as *up-down-up* by flipping a coin, with heads probability q, three times. There are eight possible sequences, corresponding to every possible way of moving from left to right along the tree. There is only one path leading to the top node at time 3, so the probability of hitting this node is $q_1 = q^3$, while the next node down can be reached in three ways, each consisting of two upwards and one downward move, so its probability is $q_2 = 3q^2(1 - q)$; similarly, $q_3 = 3q(1-q)^2$ and $q_4 = (1-q)^3$; readers may recognize this as the binomial distribution. The replication argument just outlined leads to the conclusion that the value at time 0 of an option exercised at time 3 is $V_0 = (q_1a + q_2b + q_3c + q_4e)/R^3$. This is a general fact: *the unique arbitrage-free value of an option is the expected discounted value under the risk-neutral probability of its value on exercise.* Once again: the 'probability' q is an artefact of the hedging strategy argument and is not in any sense a prediction of the real probability of an 'up' move. The real probability could be any number p with $p > 0$ and $p < 1$ and could be different at different nodes in the tree.

It is instructive to see in detail how the hedging process plays out. Suppose that $s = 100$, the per-period interest rate is 1 per cent and the up move is 10 per cent. We want to price and hedge an at-the-money (ATM) call option exercised at time 3, meaning that the strike price is equal to 100, today's price. The risk-neutral probability is $q = 0.5286$, and the option will pay 33.10, 10.00, 0, 0 at the four time-3 nodes. Using the formula above, the option value or premium is $V_0 = 8.579$, that is, 8.6 per cent of the current stock price, a fairly typical value for an ATM option.

Table 4 shows the exact sequence of events along one scenario *up-up-down*. As the option *writer* we accept the premium from the buyer in exchange for the obligation to pay the exercise value

Table 4 Option trading scenario

Time	Price move	Share price	Portf. value	N	B	Trade
0		100	8.579	0.587	−50.15	Receive premium 8.579 Borrow 50.15 Buy 0.587 shares @ 100
1	up	110	13.951	0.798	−73.82	Buy 0.211 shares @ 110
2	up	121	21.990	1.000	−99.01	Buy 0.202 shares @ 121
3	down	110	10.000	1.000	−100.00	Sell 1 share @ 110 Pay 100 to Bank Pay exercise value 10

at the end. By doing the backwards calculation described above we can figure out what the number of shares N and cash amount B have to be at each node in the tree. At time 0, $N = 0.587$ and we buy that number of shares at the current price 100, borrowing money from the bank to do so. At time 1 we experience an upward price move, and find that the value of the portfolio is exactly the option value at that node. However, the portfolio needs rebalancing, since theory dictates that we should be holding 0.798 shares, so we have to buy an extra 0.211 shares (at 110), which we do by borrowing more money from the bank. This is a *self-financing* transaction in that the total portfolio value is the same before and after the transaction. The process now continues for two more steps, and at the end we find that the portfolio value is 110 and we owe the bank 100, while the exercise value $110 - 100 = 10$ must be paid to the buyer, so we have exactly the funds required to meet our obligations.

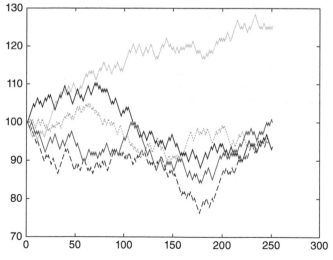

13. Binomial paths.

In summary, the option premium provides us with the funds we need to form a self-financing portfolio whose value at the end will be exactly the amount needed to pay the buyer, whatever the price moves along the way. It is irrelevant what the actual probabilities of such moves are. This replication strategy is often referred to as *perfect hedging*.

The binomial model begins to look more realistic if one considers longer time periods. Figure 13 shows five simulations of daily prices over one year (a year has 252 trading days: fifty-two five-day weeks less eight bank holidays).

Continuous-time models

It is advantageous to consider, as Bachelier did, modelling price movements in continuous time. We then obtain a *formula*—the BS formula—for the option price rather than the discrete-time algorithm of the previous section. A fundamental role is played by Brownian Motion which, we recall, is a random movement or stochastic process B_t defined for $t \geq 0$ having the properties that (i) $B_0 = 0$, (ii) it has independent increments: for any times $s < t \leq u < v$ the increments $B_t - B_s$ and $B_v - B_u$ are independent random variables, and (iii) the distribution of $B_t - B_s$ is normal with mean 0 and variance $t - s$. Wiener's 1923 paper showed that is is possible to give a rigorous mathematical construction for Brownian motion in such a way that each sample path is a *continuous function*. These functions are, however, extremely erratic, which was already evident from Einstein's observation that the 'Brownian particle' would have a mean speed inversely proportional to the square root of the time. While stock prices are not particles, we are still obliged to take account of the erratic nature of the paths because *the usual rules of calculus do not apply*. In 1944, Kiyosi Itô in Japan identified the modifications required to Newtonian calculus to make it applicable to Brownian paths, initiating the famous *Itô calculus*.

A further feature of Brownian motion is the *martingale* property. This is analogous to a player's fortune in a game of chance with fair odds, as defined in Chapter 2. Suppose you flip a fair coin several times and at each flip you win £1 on heads and lose £1 on tails. If at some point your fortune is £5 then your expected fortune after further flips is still £5 since each flip is a fair game. A martingale generalizes this property. It is a stochastic process M_t such that the expectation of M_{t+s} given all information up to t is equal to the current value M_t. This 'information' will include the past of M itself but might involve other data as well. It is clear that Brownian motion is a martingale when only its past is observed, since $(B_{t+s} - B_t)$ is independent of the past up to t and has mean zero. It is not the case, however, that every martingale has independent increments.

Bachelier used Brownian motion as a stock price model. Samuelson was dissatisfied with this choice because Brownian motion has positive probability of taking negative values while, because of limited liability, stock prices can never go below zero. For this reason he proposed taking the exponential of Brownian motion, $\tilde{S}_t = S_0 \exp(\sigma B_t)$ for some 'volatility' parameter σ, as the model, where S_0 is the current price. Using the normal density function it can be shown that the expected value of $\tilde{S}_t$ is $S_0 \exp\left(\frac{1}{2}\sigma^2 t\right)$, leading to the final stock price model

$$S_t = S_0 \exp\left\{\left(m - \frac{1}{2}\sigma^2\right)t + \sigma B_t\right\}. \tag{10}$$

When $m = 0$ it is not hard to check that S_t is a martingale, and in this case the process is known as *geometric Brownian motion*. The parameter m is the 'drift' or 'growth rate' – the expected value of S_t is $S_0 \exp(mt)$ and indeed when $\sigma = 0$ the model reduces to $S_t = S_0 e^{mt}$, equivalent to a 'bank account' paying interest at rate m. For two times $t < u$ we see from (10) that $\log(S_u/S_t) = \text{constant} + B_u - B_t$. We say that Samuelson's model has *normally distributed log returns* (note that the log-return is very close to the usual concept of return $(S_u - S_t)/S_t$ when the price increment is small), as opposed to Bachelier's *normally distributed prices*. The volatility parameter σ is then the standard deviation of log-returns over a unit length interval of time.

Throughout this book we will encounter several more complicated price models, but all of them have the common feature that they are in some sense a combination of 'drift' and 'diffusion', the diffusion being the random motion around the drift controlled by the volatility. In the present case the drift and volatility are just constants m and σ, but in other cases they may themselves be random.

To see the connection between the model (10) and the Cox–Ross–Rubinstein binomial model, note that $\log S_t = \log S_0 + (m - \sigma^2/2)t + \sigma B_t$, a normal random variable with mean $(m - \sigma^2/2)t$ and variance $\sigma^2 t$. Now consider the binomial tree as illustrated in Figure 12 with $s = S_0$ and suppose that at each fork the price—call it Y_k—moves up with probability p and down with probability $(1 - p)$. Then you can see that after n steps we have $\log Y_n = (\log S_0) + Z_1 + Z_2 + \ldots + Z_n$ where the Z_k are i.i.d. random variables each taking the value $\log u$ with probability p and $\log d$ with probability $(1 - p)$. The binomial model is determined by two parameters, u and p, and we can choose these such that the mean and variance of $\log Y_n$ match those of $\log S_t$. We can now appeal to the CLT (see Chapter 2) to show that if we fix t and increase n, at each increase redefining u, p to maintain fixed mean and variance for $\log Y_n$, then for large n the distribution of $\log Y_n$ is approximately normal, the same distribution as $\log S_t$ in Samuelson's continuous-time model, with the same parameters. This explains the utility of the binomial model as a computational method for pricing options in continuous time, which is what Cox et al. had in mind when they introduced it. (In practice, most people would use the trinomial model for computational purposes, where similar results apply but the parametrization is more flexible.)

Now for option pricing in continuous time, concentrating on a call option with strike K and exercise time T. At that time, its exercise value is $[S_T - K]^+$. The stock S_t on which the option is written is known as the 'underlying asset', or just 'underlying'. Black, Scholes, and Merton showed that the same results as in the binomial model apply here, namely

A1:   The option exercise value can be replicated by trading the underlying in the market (i.e. there is 'perfect hedging').

A2:   The option has a unique arbitrage-free value, equal to the initial investment in the hedging portfolio.

A3:   The value can be expressed as the discounted expectation of the exercise value under the unique risk-neutral distribution.

As before, we assume that in addition to the 'risky asset' S_t there is a bank account paying continuously compounding interest at rate r, so that £1 deposited at time s is worth £$\exp(r(t - s))$ at a later time t. We also assume 'frictionless trading': at each time there is a unique price S_t at which arbitrary amounts of the underlying can be bought or sold (not just integer numbers of shares). Of course, no real market is frictionless: there is always a bid-ask spread. Note from the first point above that options are in principle redundant if their value can be replicated by trading in the market. In real markets, options are not redundant because of trading costs. The theory of option pricing with trading costs is complicated, messy, and inconclusive, and the frictionless paradigm produces reliable results when trading costs are small, which in large liquid markets they are.

To introduce self-financing trading strategies the place to start is a sequential buy-and-hold strategy. Suppose we fix a sequence of times $0 = t_0 < t_1 < \ldots t_n = T$, and start with an initial endowment

£V_0. At time t_k we choose to invest in a_k units of S_t, where a_i may depend on the evolution of prices up to t_k, and sell at t_{k+1}. Then our *gain from trade* is $a_k(S_{t_{k+1}} - S_{t_k})$, the value we receive at t_{k+1} minus the value we paid at t_k. If V_k is the total portfolio value at t_k (before doing any transactions at that time) then, given that we plan to spend $a_k S_{t_k}$ on shares, our balance at the bank must be the remainder, $V_k - a_k S_{t_k}$. This shows that the gain from trade in our portfolio evolves as follows, where $R_k = \exp[r(t_{k+1} - t_k)]$:

$$V_{k+1} - V_k = a_k(S_{t_{k+1}} - S_{t_k}) + (V_k - a_k S_{t_k})(R_k - 1).$$

We leave it to the reader to check that if we define *discounted* portfolio and stock price variables $\mathcal{V}_k = e^{-rt_k} V_k$ and $\mathcal{S}_k = e^{-rt_k} S_{t_k}$ then we get the exceptionally simple expression

$$\mathcal{V}_n = V_0 + \sum_{k=0}^{n-1} a_k(\mathcal{S}_{k+1} - \mathcal{S}_k), \tag{11}$$

giving us very directly the evolution of the discounted portfolio value corresponding to a strategy (a_k). Note further that (i) the expression (11) looks like the standard discrete approximation to an integral, and (ii) the process $\mathcal{S}_k$ is a martingale (i.e. the increment has expected value 0) when m, the drift parameter in the price model, is equal to r, the bank interest rate. This is where Itô calculus enters the picture: Itô's theory of stochastic integration shows that it is possible to define integrals, denoted $\int_0^t \mathfrak{a}(s) d\mathcal{S}(s)$, which are limits of discrete-time approximations (11), where $a_k = \mathfrak{a}(t_k)$. In financial terms, $\mathfrak{a}(s)$ is the number of units of the underlying in the portfolio at time s, a hedging strategy in continuous time. Black, Scholes, and Merton showed that, for a call option on an asset with price model (10), statements A1–A3 above are true: the option can be replicated, or perfectly hedged, using a trading strategy as just described. In Property A3, the risk-neutral distribution is the distribution corresponding to the model (10) with $m = r$. One consequence of this is that the option value does not depend on the drift parameter m, and

Mathematical Finance

this is extremely fortunate as this parameter is practically impossible to estimate from data. Indeed, it has been shown that even if the data were exactly generated by the model it would take around 1,500 years to get a good estimate for m from observations of the price process.

The price of a call option price on a dividend-paying asset S_t is given by the BS formula, which has been described as 'the most important formula in financial economics'. It is:

$$BS(S_0, K, T, r, q, \sigma) = e^{-qT} S_0 \mathcal{N}(d_1) - e^{-rT} K \mathcal{N}\left(d_1 - \sigma\sqrt{T}\right)$$

where $\mathcal{N}(x)$ is the standard normal distribution function (mean 0 and variance 1), and

$$d_1 = \frac{\log(S_0/K) + \left(r - q + \frac{1}{2}\sigma^2\right) T}{\sigma\sqrt{T}}.$$

The six parameters in the formula are S_0, today's underlying asset price; (K, T), the contract specification (strike price and exercise time); r the interest rate; q the 'dividend yield' (a representation of the dividends paid by the asset); and the volatility σ. The first five are known market data but σ has somehow to be estimated. It is not obvious from the formula, but true, that the BS price increases with increasing volatility so, unsurprisingly, the option premium is high when there is lots of volatility. The corresponding hedging strategy $a(t)$ is the derivative with respect to the first (price) argument of the BS function evaluated with t as the starting time, that is, $BS(S_t, K, T - t, r, q, \sigma)$. This is known as the BS Delta and is equal to $\Delta(t) = \mathcal{N}(d_1^t)$ where d_1^t is d_1 as above but with S_0, T replaced by S_t, $T - t$. It is very natural that 'delta-hedging' is the appropriate strategy: if the underlying asset price changes by a small amount ΔS then to first order the change in option value is $\Delta(t) \cdot \Delta S$, so the delta-hedging strategy is tracking changes in option value. On the other hand it is far from obvious that the option value is given by the BS formula.

Does all this work in practice? There are many reasons why it might not. To start with, Samuelson's geometric Brownian motion model is not a particularly accurate model of real stock prices. They have 'heavier tails', meaning that the probability of large price moves is greater than that predicted by the normal distribution. Also, the volatility is certainly not a constant as it is in the BS formula. However, in the face of these objections the BS formula fights back. It can be shown mathematically that the delta-hedging strategy works, in the sense of providing the option writer with sufficient funds to pay her obligations at exercise, as long as the initial premium paid corresponds to a BS volatility that is *at least as big* as the realized volatility over the life of the option, even if the price generating mechanism differs substantially from the Samuelson model on which the BS formula is based. Given that the BS price increases with volatility this is saying that the option writer needs to quote a premium that corresponds to the highest level of average volatility that she expects over the life of the option contract. Without this mathematical result, the BS formula would indeed be an academic curiosity.

Option hedging in practice is as much an art as a science. Options are rarely hedged individually, but are bundled up in a 'book' of dozens or hundreds of contracts (in the simplest case, on the same underlying asset, say a stock index). The hedge parameters, which can be estimated, are the book Delta, as described above, and the Gamma (γ) which is the *second* derivative of the value with respect to changes in the underlying or—which is the same thing—the rate of change of Delta with respect to changes in the underlying. A perfectly hedged book (containing both the options being hedged and the constituents of the hedging portfolio) is 'Delta neutral', that is, has $\Delta = 0$, making its value insensitive to small movements in the underlying. A small Gamma means that the book will still be approximately Delta neutral even if the underlying moves a relatively long way, while if the Gamma is large, small moves can make the book unbalanced. Step 1 in constructing a well-hedged book is to choose the constituents, wherever possible, in such a way that it hedges itself, that is, both the Delta and Gamma are small. Having done that, a conventional approach is to wait until the Delta reaches some maximum acceptable level and then revise the delta-hedge to bring it back to zero. If Gamma is large, that will mean very frequent trading, with its attendant costs. Gamma can only be reduced by introducing traded options into the hedge portfolio. Trading costs for options are much higher than those of the index futures used for delta hedging, so the usual approach is to trade options only occasionally, to reset the Gamma, and then delta-hedge the whole book at a higher trading frequency. A further hedge parameter is the so-called Vega (which is not the name of any Greek letter); this is the sensitivity to changes in the volatility σ. Again, if the Vega is excessive then the only remedy is to introduce further options with offsetting Vegas. One role of the quants is to optimize this whole procedure.

An important application of the BS formula is to the pricing of FX options. For example, the USD–GBP exchange rate f_t known as 'cable' is the price of GBP1 in USD; about 1.32 at the time of writing. An FX call option would be the option to buy GBP1 for K at a time T. Its exercise value is $[f_T - K]^+$. If cable is modelled as in BS then it turns out that the option value is given by the BS formula in which the interest rate r is set at the USD rate (the 'domestic' rate, in this context) and the dividend yield q is set at the GBP interest rate (the 'foreign' rate).

The volatility surface

We now take a closer look at volatility, assumed constant in the BS theory but in practice far from constant. Aside from FX, most options are exchange-traded through organized option exchanges in major financial centres such as Chicago, New York, London, Tokyo, etc. As in all exchanges, prices are ultimately fixed by supply and demand, not by plugging some numbers into a formula. Studies of volatility in option trading are centred around the idea of *implied volatility*. First, there is a simpler and better way to write the BS formula, which is

$$BS(D, F, m, y) = DFf(y, m),$$

where $f(y, m) = \mathcal{N}(d) - m\mathcal{N}(d - y)$ with $d = y/2 - \log m/y$. The parameters here are defined by $D = e^{-rT}$, $F = e^{(r-q)T}$, $m = K/F$ and $y = \sigma\sqrt{T}$. The first three all have clear financial meaning: D is the time-T discount factor, F is the forward price for purchase of the stock at time T, and m is the 'moneyness': the ratio of strike to forward. y is scaled volatility. We call the function f the universal BS formula (Figure 14).

For each underlying asset, for example a stock index such as the FTSE100 or the S&P500, an exchange will quote bid and ask prices $c_{ij}^\pm$ for options with a range of exercise times T_i and

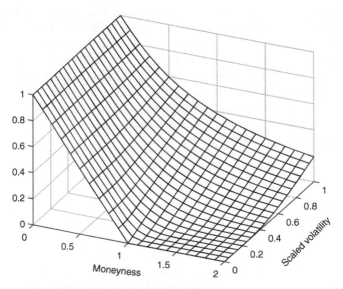

14. The universal Black–Scholes function.

strikes K_j. For purposes of volatility estimation it is usual to take the mid-price $c_{ij} = (c_{ij}^- + c_{ij}^+)/2$. For each time-strike pair (T_i, K_j) we can calculate the value of the parameter m in the universal BS formula, and then $f(y, m)$ is a function of y that increases as we increase y, so we can search over y until we find a value $\hat{y}$ such that $c_{ij}/DF = f(\hat{y}, m)$. Then $\hat{\sigma} = \hat{y}/\sqrt{T}$ is the *implied volatility* for that option. Note that the search procedure only needs to be done once, since f is a universal function. In setting up a trading operation we could start by solving for y the equation $c = f(y, m)$ for a huge range of (c, m) and storing the results in a table. Then for a given option m is calculated and the implied volatility obtained by looking it up in the table (with a little interpolation if necessary). Replacing all the given prices by their implied volatilities gives the *volatility surface*; an example is shown in Figure 15.

If prices were generated by the BS model with volatility σ then the volatility surface would be flat with all implied volatilties equal to σ, so curvature of the surface is evidence that the BS model is

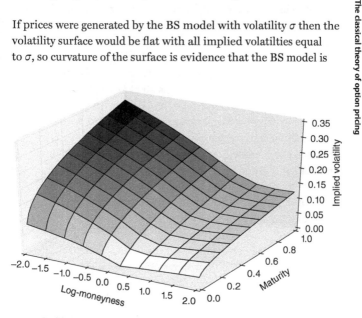

15. **Volatility surface, S&P call options: left axis is log-moneyness; right axis is time to expiry.**

incorrect for that market. Figure 15 is typical of equity markets generally, and shows that the implied volatility exhibits what is known as a 'smile': it is smallest for ATM options (those with moneyness $m = 1$, strike equal to forward) and increases as the moneyness moves away from 1 in either direction. The effect is most pronounced for short maturity options. A major topic in options research is to develop modifications or alternatives to the BS model whose volatility surfaces match the empirically observed ones more closely. The simplest approach is the 'local volatility' model in which the constant σ in the BS model (10) is replaced by a function $\sigma(t, S_t)$, so the volatility depends on the level of the underlying, as well as on the time. In an elegant analysis, Bruno Dupire showed that if we are given the whole volatility surface, that is, the value of $\hat{\sigma}(T, K)$ for all T, K then it is possible to calculate a local volatility function $\sigma(t, s)$ which, when used in the local volatility model, produces a volatility surface that *exactly matches* $\hat{\sigma}$. In reality, we only know $\hat{\sigma}$ for a finite number of pairs (T_i, K_j), but these values can be interpolated to give a complete surface (as in Figure 15). This result is widely used for risk management purposes in giving consistent prices, based on current data, for a wide range of options. In theory we could use the model to derive hedging strategies analogous to the BS Delta, but actually such strategies cannot be used because of a major limitation of the model: we derive today a volatility function $\sigma(t, s)$ for all times t beyond today, based on today's market data, but when we do the same thing tomorrow we will get a different $\sigma(t, s)$. In short, the model can only be used as a consistency check on one day, but it has no predictive power.

The way to get predictive power is to move to *stochastic volatility* models where the volatility itself is random. Typically, the volatility is represented as $\sigma(t, Y_t)$ where Y_t is an auxiliary random process. The best-known model in this class is the so-called Heston model, originally proposed in 1993, in which $\sigma(t, Y_t) = \sqrt{Y_t}$, so that Y_t is the incremental *variance* σ^2. This means that Y_t has to be positive (so we can take its square root) and, in the interests of realism, it should have a variance that is always less than some maximum level. A suitable model for this purpose, and the one chosen by Heston, is the CIR (Cox–Ingersoll–Ross) model, one of the early models in interest rate theory (see Chapter 4). The main reason for the popularity of the Heston model is that there is an efficient algorithm for computing option prices (based on the Fourier Transform). The volatility surfaces it produces are not perfect but good enough for many purposes. Much more recently, a new direction of modelling has been introduced, based on so-called *rough volatility*. Some discussion of this will be found in Chapter 8.

American options

The options we have considered so far can be exercised only at one contractually fixed time T. Such options are called *European*. Many options contracts, however, allow for exercise *at any time* up to or at a final maturity time T, and these are called *American options*. These terms were coined by Paul Samuelson who, being American, thought that this was the right name for the more sophisticated one. There is asymmetry between the two parties in an American option transaction. The buyer (holder) of the option has the right to exercise at any time. The seller (writer) has the obligation to pay the exercise value *whenever* the buyer chooses to exercise. There are a couple of things we can say right away. First, the American option cannot be worth less than the corresponding European one, because waiting until T and then exercising is one possible strategy for the buyer. The buyer has extra 'optionality', and this may be worth money. Second, it turns out that in the case of a call option written on a non-dividend-paying asset the European and American values are the same: the best strategy for the American holder is to exercise at T, so in this case the extra optionality has no value. This is a general fact, not restricted to the BS model. For this reason, studies of American options focus on the put option, where there is extra optionality even with no dividends.

It is fairly clear in general terms what the holder of an American put should do. The exercise value at t is $[K - S_t]^+$, paid at time t so the holder should wait until the current price S_t is some distance below the strike K. There is a trade-off between early exercise—take what's on offer now, it's worth less if taken later—and waiting for a lower price that might come along. 'Take it now' is increasingly attractive as the maturity time approaches. The optimal strategy is typically as shown in Figure 16 where the curved line is the 'exercise boundary' and the holder should exercise when the price trajectory hits this boundary. The strike is 10 and the option is never exercised when S_t exceeds this level. The boundary rises rapidly up to the strike as maturity approaches and the chance to get more by waiting disappears.

The mathematics behind all this is sophisticated. The value of the option in the holding region satisfies the same partial differential

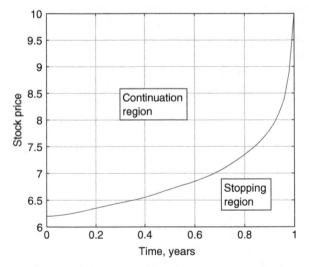

16. **American put option exercise boundary (maturity = 1 year; strike = 10).**

equation as in the European BS case, but we have to determine where the boundary between the holding and exercise regions lies. This is determined by the requirement that the expected discounted value of the option payoff should be maximized when the option is exercised at the first boundary hitting time. All this was worked out in 1965 by MIT mathematician Henry McKean as an appendix to a paper by Samuelson. McKean was a key figure in the development of stochastic calculus and indeed had collaborated directly with Itô. However, 1965 was pre-BS and it was only in the 1980s that the complete arbitrage theory of American options was finalized, by A. Bensoussan and I. Karatzas.

For American options there is no pricing formula similar to the BS formula for the European case. Solutions have to be obtained computationally. There is a very simple solution when the price process is represented by a binomial tree, and this is one of the reasons why the introduction of the binomial tree was such a breakthrough. In the tree, one starts at the end, where exercise values are known, and works backwards. At any node N in the tree, the option value is the discounted average under risk-neutral probability q of the values at the two nodes to which the price may move from N. Call that the 'continuation value' at N. The 'immediate exercise value' at N is $[K - S]^+$ for the put option, where S is the underlying price at N. Now replace the continuation value at N, and all other nodes at the same time step, by the immediate exercise value if the latter is bigger, and carry on working backwards. This simple modification to the algorithm produces the American value at the initial time. The nodes where immediate exercise is better define the exercise region. As we pointed out earlier, by constructing binomial trees with more and more smaller and smaller steps we can approximate geometric Brownian motion and hence the BS American price. In practice, the binomial tree is replaced by a trinomial tree, using just the same algorithm, because of the extra flexibility in calibration. Figure 16 was produced this way.

The Fundamental Theorem of Asset Pricing

So far we have looked at various cases. In the binomial tree, there is a unique choice of risk-neutral probability q such that an option with payoff G can be perfectly hedged if priced at its discounted risk-neutral expectation $\mathbb{E}_q[G/R]$. At any other value there is an arbitrage opportunity, by the Law of One Price. In a trinomial tree there are many possible choices for the risk-neutral transition probabilities and hence no unique option price. Uniqueness can be recovered by specifying a market price for one traded option, but in general it is true that an option creates arbitrage unless its price is the discounted expectation under *some* risk-neutral probability. No probabilities are required to specify the model. Moving on to the BS model, probabilities are required since the model is based on Brownian motion. However, by changing the original drift from m to r, the riskless interest rate, the discounted price is a martingale and an option can be perfectly hedged when priced at its risk-neutral expected value, so once again there is arbitrage unless it is valued that way.

Is all of this part of some 'big picture' that would also include wider classes of underlying price models such as discrete-time processes taking arbitrary values rather than just a finite number of values (as in binomial or trinomial trees), or more complicated continuous-time processes? If so, there ought to be a general theorem stating that a model does not present arbitrage opportunities if and only if the discounted price process is a martingale under some modification of the original probabilities. Such a result would be called the FTAP. The appearance of the BS formula set off a twenty-year quest for the FTAP, the final version being hunted down in two famous papers published in 1994 and 1998 by Delbaen and Schachermayer. FTAP is true but only after subtle modifications in the exact definition of 'arbitrage' and 'risk-neutral distribution'. We cannot go into details here, but can outline one or two of the ingredients.

A classic arbitrage strategy is 'doubling'. Suppose a coin is flipped repeatedly and you win or lose your stake at each turn. If you double your stake on each loss you are bound eventually to win. For example, if the initial stake is 1 and the first four throws are losses then you will have bet and lost $1 + 2 + 4 + 8 = 15$, but if the next bet 16 is a win then all your losses are erased and you have a net gain of 1. This strategy has long been called a martingale—it appears under that name in Thackeray's *Vanity Fair* (1848) along with a lurid description of the dire fates of the English beaux who played it at the gambling tables of Baden Baden. The catch is of course that the losses pile up alarmingly fast and the player is quite likely to run out of money before that long-awaited win saves the day. One can show (using martingale theory) that there can be no arbitrage strategy if the player's losses are never allowed to exceed some given limit. So any set-up where an unlimited number of bets can be made, a lower limit on the value of hedging strategies must be imposed.

Next, suppose we were to propose a stock price model such that the price paths are continuously differentiable, as shown in Figure 17. This means that at any time such as t_1 the path has a tangent, as shown, and if the slope is strictly positive then it will still be positive for some time beyond t_1, so there will always be some time $t_2 > t_1$ such that $S_{t_2} > S_{t_1}$. As traders, all we have to do is buy at t_1

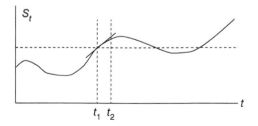

17. **Continuously differentiable price path.**

and sell at t_2 to make an arbitrage profit. To get a sensible model for finance, this must be ruled out, so if we want a model with continuous price paths, those paths cannot be continuously differentiable. Wiener proved that the Brownian motion path is *nowhere differentiable*, so this kind of arbitrage cannot occur in any model, such as the BS model, which is driven by Brownian motion. This is one of the many respects in which stochastic calculus seems miraculously tailored to the needs of financial modelling. Later authors showed that the nowhere differentiable property holds for all continuous-path martingales, so these processes are all viable candidates for use as asset price models, explaining the appearance of martingales in the FTAP.

Chapter 4
Interest rates

The world of interest rates is where Keynes's comment about money being a link between the present and the future is most apposite. Interest-rate related contracts can stretch out to thirty years or even longer, with high volumes of trading in the five to ten year range, giving an informative peek into the future.

In ordinary life one comes across many different interest rates—mortgage rates, rates on credit cards, rates paid by savings accounts, etc. These are mirrored in the commercial world by loans for short-term funding and longer term loans for business development and project finance. The reasons there are so many different rates are market conditions (how competitive is the market in that particular sector?) and *credit risk* (how sure is the lender that the loan will be repaid?). All of this used to be a matter between the bank and its clients, but organization of, and trading in, credit risk has ballooned in the 21st century; this is the subject of Chapter 5. First we will discuss trading of interest rates, focussing on the inter-bank market (that is, trading between banks) which is the area where most trading takes place, where most quant effort is deployed, and in whose development mathematical finance has played a key role. It is a satisfyingly 'clean' subject in that, for any one currency, there are only a handful of traded assets in stark contrast to the equity world where there are thousands of traded assets.

To get started we need some basic terminology. Interest rates are invariably quoted in annualized terms as percentages of some principal or 'notional' amount. The interest payments or 'coupons' may however be paid more frequently, for example quarterly, in which case something closely equal to a quarter of the annualized rate will be paid on each coupon date. There are sometimes confusing 'day-count' conventions the purpose of which is to determine precisely how much will be paid. These things are the bane of the quant's existence, but irrelevant to our discussion here. We will suppose that interest rates are directly stated on a per-period basis, so at rate c the coupon paid on a principal amount M is just $M \times c$.

A *bond* is a traded security specified by its maturity date, notional amount M, and interest rate c. The holder receives the stream of coupons up to the maturity date, at which the last coupon is paid together with the notional amount. Suppose the holder has bought the bond for price v when there are n periods left to maturity. The bond then has 'yield' y, defined as the interest rate such that a bank account with initial deposit v and paying interest at that rate would generate enough money to make all the coupon and principal payments on the bond. Specifically: an initial deposit q in an account paying y will be worth $q(1+y)^k$ after k years. The bond pays coupon Mc at that time, so an initial deposit $q_k = Mc/(1+y)^k$ would be enough to finance this coupon. Now sum these amounts over all coupon and notional payments to arrive at a total $q(y)$, and choose y such that $q(y) = v$, the market price of the bond. This value of y is the yield. It turns out (check the algebra!) that $y = c$ when $v = M$, that is when the price of the bond is equal to its notional. In this case we say the bond is *at par*. Note the inverse relationship between price and yield: yields fall when prices rise. A bond is usually at or close to par when first issued, but its price in the secondary market will adjust over time to be in line with prevailing market rates. Thus if rates fall then a bond that used to be at par will be paying an above-market coupon; this is valuable, so its price must rise causing its yield to fall into line with the current market.

A *zero coupon bond* (ZCB) pays no coupons but pays £1 at its maturity date T. The price at some earlier time t is written $p(t, T)$ and directly represents the discount factor or 'time value of money': a pound at time T is worth $p(t, T) < 1$ at t and the t-forward price for delivery of a non-dividend paying asset S at time T is $S_t/p(t, T)$. ZCBs are not directly traded in the market, but conceptually are the key quantities, needed whenever we want to value a basket of contracts making payments at different times. We will see later how ZCB prices are inferred from traded assets in the market.

The key data defining the inter-bank market are short-term rates, interest-rate futures, and swap rates. The short-term rates are known as LIBOR (London Interbank Offered Rate). These are compiled, for various currencies, by ICE (Intercontinental Exchange), an independent company, and are averages of rates charged by banks for short-term loans to other banks. The rates are announced daily. The terms specified range from one day ('overnight') to one year. Banks value their assets by discounting at the LIBOR rate. It is important to note that interest rates are always *set in advance and paid in arrears*: the rate is set at the beginning of the period it relates to (so investors know what they are in for) but payment is made at the end of the period. The three-month LIBOR rate is a standard 'benchmark' of a bank's funding costs, and there is a futures market on this rate. It works in exactly the same way as the index futures market described in Chapter 1 except that the final price to which a contract settles is the current 3m LIBOR rate (times some specified notional amount). Futures contracts are available with maturity times up to two or three years ahead. They are widely used for hedging, and their prices provide additional information about current rates.

Finally, there is the swaps market. In an interest rate swap two parties exchange payments on two bonds with the same notional amount, maturity, and coupon dates. One bond pays a fixed rate of interest c while the other pays, at each coupon date, a 'floating' rate equal to the LIBOR rate set at the preceding coupon date. The two parties are identified as the 'payer' and 'receiver', the payer being the one who pays the fixed-rate coupons. Since the bonds have the same notional, the final exchange of notional at the maturity time cancels out, so the whole process only involves exchange of coupons; this explains the term 'notional': the principal itself is never in play and the notional just defines the size of the contract. Nevertheless, it is conceptually helpful to think of a swap as an exchange of bonds. It is a remarkable fact that *the value of the floating-rate bond with unit notional is 1*; remarkable because at the beginning of the deal we have no idea what the values of later LIBOR rates will be. The point here is that the bonds are valued by discounting at the LIBOR rate, the same rate as the interest they pay. If the rate for the final period is L then the payment at maturity is $(1 + L)$, the notional plus the final interest payment; discounting at LIBOR, the value of this payment at the beginning of the final period is $(1 + L)/(1 + L) = 1$. Continuing backwards, it follows that the value is 1 at any coupon date and hence at the initial time. The same argument can be used to show the result already mentioned that a fixed rate bond is at par when its yield is equal to the coupon. Since the floating-rate side of a swap is always at par, the value of the swap is determined by the value of the fixed-rate side, which is obtained by summing the values of all the payments, discounted by the ZCB value for the time at which the payment is made. The *swap rate* is the value of the fixed-side coupon c such that the swap has value zero at inception.

Swap rates are *market data*. There is a very large and liquid market in swaps. They are traded with no up-front payment, so have 'fair value' only when the fixed-side rate is what the traders perceive to be the true swap rate, which will be fixed by negotiation and 'supply and demand'. The swap rate will depend on the maturity time and, on any day, quotes will be available for swaps with maturity one year, two years, etc., up to ten years, with sparser quotes after that stretching out to thirty years. What is *not* market data is the ZCB value. If we did know the ZCB values we would be able to calculate the swap rates, but in fact the problem is the other way round: given the swap rates, can we calculate the ZCB values? Indeed we can; the algorithm that does this is called a *yield curve generator*. The result is generally expressed in terms of the zero-coupon yield, that is, the interest rate corresponding to the ZCB value. Figure 18 shows the yield curves out to ten years for three currencies on a day in October 2017. Note how low the rates still are: even for the USA, the ten-year rate is under 2.5 per cent, and the Euroland rate is actually negative at the short end, unthinkable ten years ago. This situation is discussed in Chapter 8. Figure 18 is based only on swap rates and does not cover periods of less than one year. To fill in the short end, a commercial yield curve generator will use the other rates in the market, a selection

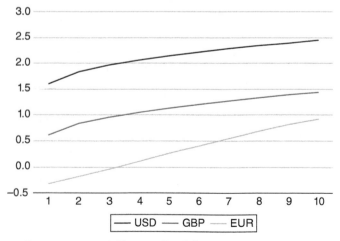

18. One- to ten-year yield curves (in %) for USD, GBP, EUR, 25 October 2017.

of LIBOR rates and futures. The algorithm will always involve interpolation, in order to determine the curve at arbitrary times, but given the amount of data available different banks will come up with yield curves that agree with each other to a high degree of accuracy. Calculating the yield curve is one of the most basic things a bank has to do.

The yield curve is a stochastic object, moving around over time. The reader is strongly recommended at this point to visit the website https://stockcharts.com/freecharts/yieldcurve.php where an animation showing the movement of the USD yield curve over the period 1999–date will be found. This movie deserves an Oscar for its dramatic twists and turns. One of the major achievements in mathematical finance has been to develop effective and elegant models to describe these movements.

The standard options contracts in interest rates are *caps* and *swaptions*. Like other options, they are only exercised if the payoff value is positive, so a premium is charged at inception for entering these contracts. If we are paying the LIBOR rate we are exposed to the risk that this might rise. Caps hedge this risk by specifying a maximum rate K, so if at a particular coupon date the LIBOR rate is L_k then the amount paid (per unit notional) is the lesser of L_k and K. Otherwise stated, the payoff of the option is $[L_k - K]^+$, so the cap is a call option on the LIBOR rate. In fact it is a strip of call options, one for each coupon date k. The individual options are called *caplets* and are priced one-by-one, the price of the cap being simply the sum of the caplet prices. Swaptions, on the other hand, do not break down into pieces in this way. A swaption is the right to enter a swap at some specified time T in the future at a fixed-side rate K set now. It is a *payer's swaption* or a *receiver's swaption* depending on whether the right conferred is the right to enter on the fixed-rate side or the floating side. A payer's swaption will be exercised only if the agreed rate K is less than the swap rate S_T that materializes at the exercise time. Then the holder is receiving the floating rate coupons but paying a below-market rate to get them. A primary function of the swaption is that it provides a mechanism for getting out of a swap. A swap is like a forward contract: it may turn out to be favourable or unfavourable for either of the parties. Suppose I am the receiver in a swap scheduled to terminate at T_2 with fixed-side rate K. Then I could negotiate a payer's swaption with the same notional to enter at some time T_1 between now and T_2 a swap extending over the period from T_1 to T_2 at the same fixed-side rate. If exercised, the second swap cancels out all remaining cash flows from the first swap, so the effect is just to cancel that swap. A more sophisticated contract is the *Bermuda swaption*. Here a notional amount and a fixed-side rate K together with a list of coupon dates up to a final terminal date T are agreed and the holder has the right to enter a swap, terminating at T, at any coupon date prior to T. The holder can then monitor the market and exercise (if at all) at the most favourable time. This is analogous to an American option, but generally an American option can be exercised at any time, whereas the Bermuda swaption can be exercised only at one of a finite list of times. The option is like Bermuda: between America and Europe, but closer to America.

Turning to stochastic modelling for interest rates, this is considerably more complicated than the modelling for stocks, indices, or FX, for two reasons. First, in the latter cases a small number n of asset prices $(S_t^1, \ldots, S_t^n)$ is being modelled, so the output of the model at any time is an n-vector. In interest rates, the object to be modelled is the yield curve, which is 'infinite dimensional' in that it cannot be represented exactly by a finite list of numbers. Second, in the case of stocks, etc., option payoffs are directly functions of the prices, while interest rate options are functions of quantities derived from the yield curve, such as LIBOR rates, which are not generally modelled directly. An exception is the LIBOR Market Model (LMM) discussed later in this chapter, and this is one of the reasons this model is so attractive. A consequence of the second point is that hedge parameters are also not direct outputs from the model and there is a variety of ways to get at them. Which should be used depends on the problem. In general, the theory of hedging in interest-rate markets is not a very well-developed subject, not because it can't be done but rather because, in practice, only quite simple strategies are implementable.

The oldest and most traditional approach to interest rate modelling is to model the *short rate*, which arises from the idea of continuous compounding. If we have a bank account starting with £1 and earning interest at a continuously compounding rate r then the balance at time t will be $\exp(rt)$. If the rate is time-varying, $r = r(t)$, then the balance is $\exp(\int_0^t r(s)ds)$. A short rate model assumes that $r(t)$ is a stochastic process, that is, evolves randomly over time. One of the best-known versions, the Hull–White model, assumes that $r(t)$ satisfies equations that are very similar to those satisfied by the log-price in BS, but with one important difference. Stock prices can rise to extreme values, while interest rates are 'mean reverting': they may sometimes rise to quite high values (but don't hold your breath), but will come back down again to levels in the range of, say, 0–10 per cent. This mean-reverting behaviour is easily obtained by a simple modification of the BS model, and then ZCB, prices can be computed in closed form. They take the form $p(t, T) = f_h(t, T, r(t))$, that is, all ZCB prices are determined by the single number $r(t)$. The model includes a time-varying parameter h which can be adjusted to a value h_0 such that today's yield curve, the prices $p(0, T)$ are indeed equal to $f_{h_0}(0, T, r(0))$; the model is 'calibrated' to the yield curve. There are two further parameters, a volatility parameter σ and a mean-reversion parameter ℓ, that can be used to calibrate the model to market prices of caps or swaptions (but not both at the same time: it is a fact of life in interest rates that the cap and swaption markets are different things and no model can simultaneously calibrate to both). The Hull–White model is efficiently implemented as a trinomial tree (see Chapter 3). It used to be frowned upon because the short rate, being normally distributed, can be negative with positive

68

probability; but now that negative rates do appear in the market (see the Euro rates in Figure 18) this feature is actually an advantage.

Short rate models have several shortcomings. The first is that they need re-calibrating every day. We choose h_0 at time 0 to match the prices $p(0, T)$, but the next day we observe a new yield curve that is not exactly matched by the model, so the parameter h_0 has to be adjusted to a new value, say h_1. This is similar to the BS model for equities where we choose an implied volatility to match option prices today but will need a different volatility tomorrow. In both case the conclusion is the same: the model is not rich enough. To take this further we need to examine the range of yield curve moves that occur in reality versus the moves the model is able to generate. The standard way to study yield curve moves empirically is to collect a sample consisting of a year or more of daily yield curves and applying a statistical technique known as Principal Components Analysis (PCA) which pulls out the main 'directions' in which yield curve moves take place. The results are shown schematically in Figure 19. The dominant move is a 'parallel shift', (a), in which the curve moves bodily up or down. The next most important move is a rotation (b) in which the moves at either end of the curve have different signs, and finally, less significantly, a

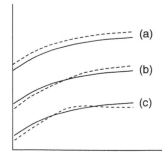

19. **Yield curve moves: (a) parallel shift, (b) twist, (c) hump. Solid line is original curve; dotted line is shifted curve.**

'parabolic' type of move (c) in which the central part of the curve moves up and the wings down, or conversely.

Returning to the short rate model, all the ZCB values, and hence yields, are functions of one parameter, the short rate $r(t)$, with the same sign: all ZCB value go down when $r(t)$ goes up. This implies that the output curve *can only be a parallel shift* of the time-0 curve. That is what is meant by saying that the short-rate model is not rich enough.

Figure 20 shows schematically what an ideal model should look like. It takes a yield curve $C(0)$ (today's curve, real data) and a noise signal to provide some randomness, and generates a random yield curve $C(t)$, say tomorrow's curve, as output; $C(t)$ is then a 'random variable'. Call the *support* $\mathfrak{S}$ of the output the smallest set of yield curves such that $C(t)$ belongs to that set with probability one; so $\mathfrak{S}$ is essentially the set of all possible curves that our stochastic model can produce. Then $\mathfrak{S}$ should be big enough so that when tomorrow's curve is revealed, it belongs to $\mathfrak{S}$; that is, the model is rich enough to encompass anything that might happen in practice. This is the test that short rate models fail: because only parallel shifts are possible, it is very likely that tomorrow's curve is one that is impossible within the model, an unsatisfactory state of affairs.

An influential approach to whole-yield models along the lines of Figure 20 was initiated by Heath, Jarrow, and Morton (HJM), and

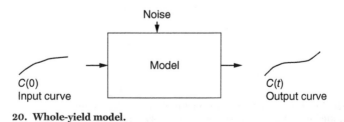

20. Whole-yield model.

independently in the UK by Babbs, in 1992; it is now universally referred to as 'HJM'. The theory is couched in terms of instantaneous forward rates, so we have to take a little detour to understand what that means.

Given times $T < U$, the ZCBs $p(0, T)$ and $p(0, U)$ define continuously compounding interest rates r_T and r_U over the periods from 0 to T, U respectively by $p(0, T) = e^{-r_T T}$. The *forward rate* r_f is the rate such that investing at r_T up to T and then switching to a rate r_f from T to U is equivalent to investing at r_U from 0 to U. It is easy to see that $r_f = (U r_U - T r_T)/(U - T)$ and $e^{r_f(U-T)} = p(0, T)/p(0, U)$, or equivalently that, with $x = U - T$, r_f is equal to $(\log p(0, T) - \log p(0, T + x))/x$. If x is small then this is an approximation to the 'infinitesimal' short rate $f(0, T)$ where $f(t, T) = (d/dT) \log p(t, T)$. By calculus, this implies that $p(t, T) = \exp(\int_t^T f(t, u) du)$. Thus there is a one-to-one relationship between $f(t, T)$ and the ZCB $p(t, T)$; the forward rate also determines the short rate, which is simply $r(t) = f(t, t)$. The reason for the choice of the forward rate as the object to model is that if there is no volatility then $f(t, T)$ has the same value $r(T)$ whatever t. That is because for any s, t we then have $\log p(s, T) = \log p(s, t) + \log p(t, T)$, so differentiating with respect to T the first term on the right drops out. A model will start with an initial condition $f(0, T)$ and, as indicated in Figure 20, specify a mechanism for producing a range of random values $f(t, T)$ at future times t. With no volatility, that is, no noise input, $f(t, T)$ does not vary at all—it is always equal to $f(0, T)$. So the purpose of the model is to describe how the deviation from this constant value evolves when there *is* volatility. As in BS, a model will always have two basic components, 'drift' and 'volatility'. The main result in HJM is that the drift is completely determined by the volatility, so only the latter needs to be specified to complete the model. How to do this has however proved to be a daunting task. The model is 'infinite-dimensional' in that we have to compute the solution for all the values of T from 0 up to some final maturity $T_{\max}$, so some quite sophisticated approximations will certainly be required when it comes to writing a computer program to implement the model. A minor industry sprang up concerned with developing appropriate theory, identifying cases where HJM is actually finite-dimensional and looking at simulation methods for doing the computations.

As so often in science, a shift in perspective broke the logjam. It turned out to be a better idea, instead of modelling instantaneous forward rates, to model *forward LIBOR rates*. Instead of HJM we get 'BGM' for the originators Brace, Gątarek, and Musiela, otherwise known as 'LMM', an equivalent construct due to Jamshidian. Suppose we have coupon dates $t_1, t_2, \ldots$ A forward LIBOR contract involves three times $t < t_k < t_{k+1}$. The relationship between the LIBOR rate set at t_k and the ZCB value is $p(t_k, t_{k+1}) = 1/(1 + L_k)$ since £1 accrues to £$(1 + L_k)$, so that £$1/(1 + L_k)$ accrues to £1. Thus the value at t_k of the payment made at t_{k+1} is $p(t_k, t_{k+1})L_k = L_k/(1 + L_k) = 1 - p(t_k, t_{k+1})$. Now suppose that at time t we form a portfolio that is long one t_k-bond and short one t_{k+1}-bond, with value $S_t = p(t, t_k) - p(t, t_{k+1})$. At t_k its value will be $S_{t_k} = 1 - p(t_k, t_{k+1})$, as the t_k-bond has matured to 1, exactly hedging the LIBOR payment. The time-t forward value of S_{t_k} is $f_t = S_t/p(t, t_k) = 1 - p(t, t_{k+1})/p(t, t_k)$ and the forward LIBOR rate $\mathcal{L}_k(t)$ is this value translated into 'rate' terms $f_t = \mathcal{L}_k(t)/(1 + \mathcal{L}_k(t))$, giving $\mathcal{L}_k(t) = p(t, t_k)/p(t, t_{k+1}) - 1$. Note that $1 + \mathcal{L}_k(t)$ is a ratio of bond prices, which can be interpreted as the value of the t_k bond expressed in units of the t_{k+1} bond as 'numéraire' asset. The BGM and LMM models make full use of this flexibility in the choice of a unit of account.

The LMM is a stochastic model representing the movements of a finite collection of forward LIBOR rates $\mathcal{L}_0, \mathcal{L}_{t_1}, \ldots, \mathcal{L}_{t_{m-1}}$ where 0, $t_1, \ldots, t_m$ are the coupon dates for, say, a cap initiated at time 0 and maturing at t_m. Note that the initial values $\mathcal{L}_k(0)$ are determined by the yield curve at the initial time and that $\mathcal{L}_0(0) = L_0$, the LIBOR rate set at that time, so no modelling of $\mathcal{L}_0$ is required, and each $\mathcal{L}_k(t)$ is 'alive' for times t between 0 and t_k where the forward rate settles at $\mathcal{L}_k(t_k) = L_k$. There is no randomness beyond t_{m-1} when the last LIBOR rate is set. Associated with each component $\mathcal{L}_k$ is a Brownian motion W_t^k (possibly correlated with other W_t^j) together with drift and diffusion coefficients. As in HJM, the nub of the matter lies in choosing the volatility functions. They are calibrated to whatever market cap rates are available for maturities in the range 0 to t_m. The whole model is simpler than HJM in that it is not infinite-dimensional; there are only $m - 1$ components. That could be a large number, though; for a five-year cap with quarterly coupons $m - 1 = 19$, so a considerable amount of computational effort could be required to obtain the solution. As in HJM, the drift coefficients are determined by the volatilities. Jamshidian's derivation of the formula that relates the two is one of the most elegant in the mathematical finance literature, involving an ingenious juggling of different numéraires (units of account) for the different forward LIBORs. A further pleasant property of the LMM is that when the volatilities are deterministic (i.e. non-random) then caplet prices are given exactly by a version of the BS formula.

All of the models considered can be used to calculate prices of the standard interest-rate derivatives, caps, and swaptions, in European or Bermuda form, although, as remarked earlier, separate calibration is needed for the two markets. In view of its complexity, Monte Carlo is the only viable computational technique for the LMM. However, this application required new developments in Monte Carlo. Recall from Chapter 3 that in computing the American option value we work *backwards* from the known exercise value at the final maturity time. Monte Carlo, on the other hand is essentially a *forwards* algorithm starting with initial prices and simulating future trajectories. It was a significant challenge to come up with Monte Carlo based algorithms that could incorporate the early exercise feature.

A major use of swaps, and the reason for the huge size of the market in them, is as hedging instruments for these derivatives.

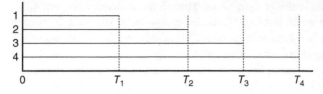

21. Maturity times for four hedging swaps.

The value of, say, a swaption at any time depends on the yield curve observed at that time covering the period up to the swaption maturity. Readily traded swaps depend on the same thing and so can be used to delta-hedge the swaption.

The procedure in brief outline is as follows, referring to Figure 21 in which T_4 is the maturity date of the swaption and we want to set up an initial hedge at time 0. Our hedging instruments are four swaps with maturities as shown in the figure. Each of the swaps depends on the yield curve up to its maturity, but not beyond, so we start with the T_4 and calculate the sensitivity of its value, and the value of the swaption, to F_3, the forward swap rate at T_3. By taking the ratio and trading that number of T_4 swaps we form a portfolio that is insensitive to movements in F_3. Moving backwards to T_2, we calculate the sensitivity of that portfolio to movements in F_2 and cancel out that sensitivity by adding the right number of T_2 swaps. Continuing in this way we arrive at a portfolio that is insensitive to any of the forward swap rates. Set up at 0, the portfolio will of course need rebalancing, with further trading in the swaps, as time goes on. Of course, there is much more to swaption hedging than this, but delta-hedging is the most basic ingredient.

Finally, we should mention the disruption of the interest rate market occasioned by the 2008 crisis. It used to be a beautiful world where there was one yield curve per currency and modelling was dominated by exceptionally elegant constructs such as the

LMM. Many relationships between various quantities follow, one of which concerns forward LIBOR rates. At any time t we can observe the three- and six-month LIBOR rates L_3 and L_6. We know the value of L_6 now, but we could also secure a LIBOR-based payment at six months by entering a forward rate agreement (FRA) for L_3 set three months from now. We should not be able to make an arbitrage profit by transactions of that kind, and this implies some relationship between the three- and six-month rates. The aftermath of the crisis saw a breakdown of trust between banks, so that a promised payment six months from now is regarded as more risky than one after three months, that is, *credit risk* has entered the picture. This has led to a world of multiple yield curves—one for each payment frequency—making valuation problems substantially more difficult. It means that the world of interest rates is no longer insulated from the world of credit, the subject of Chapter 5.

Aside from the credit question, LIBOR came under scrutiny in the aftermath of the crisis when it became clear that some traders in major banks had been colluding to manipulate the rate. This led to a general revision and tightening of the procedures and also to a search for an alternative to LIBOR as a benchmark interbank rate. There are several contenders but at the time of writing no final decision. It is a complex question if only because of the huge volume of live LIBOR-based contracts, all of which would have to be renegotiated were LIBOR not to be retained.

Chapter 5
Credit risk

Bonds were briefly described in Chapter 1. They are issued by corporations or other entities wishing to raise funds without creating new equity. Of course commercial loans have been around for centuries—see Shakespeare's *Merchant of Venice*—but the point about a bond in the modern sense is that it is a tradable asset. The issuer receives the principal amount on the issue date, then pays (to whoever owns the bond at the time) interest coupons at regular intervals and finally repays the principal along with the last interest coupon at the maturity date. Given this structure, what determines the coupon rate? It must be related to the 'credit worthiness' of the issuer, and this is formalized by rating agencies, of whom there are a small number operating globally, the best known two being Standard and Poor's (S&P) and Moody's. They assign a credit rating to each company under consideration, on a scale which, in the case of S&P for example, reads AAA, AA,... as listed in the first column of Table 5, AAA ('triple-A') being the most secure. A company's rating is based on a careful analysis of its accounts, its trading position, its performance in the stock market, and any other factors deemed relevant. The borrowing cost, the rate a company will have to pay to issue a bond, is closely related to its credit rating. It will never be less than the current rate offered on government bonds of comparable maturity, since the latter are deemed risk-free, and won't be much more than that for a triple-A corporation; the excess increases quite rapidly as we

Table 5 S&P Global Rating Transition Matrix 2016

From/to	AAA	AA	A	BBB	BB	B	CCC/C	D	NR
AAA	81.25	12.50	0.00	0.00	0.00	0.00	0.00	0.00	6.25
AA	0.00	90.11	6.50	0.00	0.00	0.00	0.00	0.00	3.39
A	0.00	0.82	91.02	4.01	0.00	0.00	0.00	0.00	4.15
BBB	0.00	0.00	2.62	87.70	3.23	0.17	0.06	0.00	6.23
BB	0.00	0.00	0.00	3.12	80.37	6.07	0.23	0.47	9.74
B	0.00	0.00	0.00	0.00	3.84	74.00	5.55	3.68	12.92
CCC/C	0.00	0.00	0.00	0.00	0.99	14.36	40.59	32.67	11.39

go down the rating order. The actual value will be fixed only days away from the issue date, depending on market demand. Generally bonds are at or very close to par at the issue date, that is, a £100 bond is sold for £100, but over time prices in the secondary market will adjust so that the yield on bonds with the same credit rating are closely aligned.

Credit ratings are revised at regular intervals and what is happening in the market is revealed by so-called 'credit transition matrices' such as the one at Table 5 which refers to the calendar year 2016. To construct it, list all the companies that were rated (say) AA on 1 January 2016, then list the ratings of the same companies on 1 January 2017. The numbers in the table at row AA are the fraction (in %) of the AA companies that migrated to the ratings given in the columns. We see that 90.11 per cent of AA companies stayed AA, while 6.5 per cent were downgraded to A. The last two columns are D for default, meaning that the company has gone into bankruptcy, and NR for not rated, meaning that the company is no longer rated, usually because it has merged with, or been taken over by, another company. In 2016, no AA firms defaulted and 3.39 per cent were no longer rated by the end of the year. It turns out that the numbers in the table are quite stable over successive years. Some analysts prefer to average them over, say, five years to get a more stable picture.

It is striking to realize that, while in 1980 there were sixty AAA-rated corporations in the USA, in 2018 just two remained, Microsoft and Johnson & Johnson; the third contender Exxon Mobil was downgraded in 2016 for the first time since 1949 due to instability in the oil market.

Turning to CDS, these were already outlined in Chapter 1. A CDS contract is written on a *specific bond* issued by XYZ plc, not on XYZ plc as a whole, and the concept of default is very different from default of a rated company as discussed earlier. For a CDS, 'default' means any infringement, however technical, of the terms of the bond contract, which provide for payment of exact amounts of coupon and principal on specific dates. XYZ plc could certainly default in this sense by some mischance while still being very much a going concern. When a default occurs the CDS writer (or 'protection seller') ABC has to pay the holder (protection buyer) the difference between the face value of the bond and its 'recovery value' which essentially is its post-default market value. This means that, in broad terms, holding an XYZ bond plus a CDS written on that bond is equivalent to holding a risk-free bond, since either there is no default or else the CDS holder will receive at default an amount equal to the face value of the bond, so the effect in economic terms is simply that the bond has matured early. For this reason the CDS premium paid by the protection buyer must be close to the 'spread' on the underlying bond, the spread being the amount by which the bond coupon exceeds that of a risk-free government bond.

Credit risk modelling and analysis

There are two risks in writing or holding a bond or CDS: market risk and credit risk. The former concerns the relationship between the coupon paid and current market conditions. Bond prices in the secondary market must adjust so that yields are comparable to current levels in the market. For example, suppose a bond is issued when rates are high and consequently pays a large (fixed) coupon. If rates then fall, the bond is 'off-market'—it is paying too much—so investors will pile in, pushing the price up above par and the yield down until the yield matches current market levels. Analysis of this phenomenon is a part of interest rate theory as discussed in Chapter 4 and is mainly a question of market volatility. Credit risk is a very different story concerned with the modelling of times of default, which are discrete events, not directly related to volatility. In the case of a CDS for example, a model should enable us to evaluate, at any time, the probability of default at subsequent coupon dates or at maturity.

As for other asset classes, credit risk models are required in two varieties, 'real-world' models that statistically describe the evolution of prices as they are observed in the market, and 'risk-neutral' models used for pricing à la BS. Often, the primary model will be the risk-neutral one, the 'real-world' model being the same but with an extra term reflecting the so-called *market price of risk*.

Credit risk models are divided into two main categories, known as 'structural form' and 'reduced form'. The former was initiated—as were so many things in mathematical finance—by Robert Merton, in a 1974 paper whose influence lives on. It is concerned with corporate debt and instantiates the idea that a firm will default when its assets are insufficient to cover a liability when the latter becomes due. Figure 22 shows this idea in the case of a single liability to be paid at time T; it shows two possible evolutions of the value of the firm's assets V_t, the solid line resulting in default as the end value is less than K, the liability. It is apparent from the figure that the default risk is a *put option on the value of the firm*: at time T the amount of the liability that the firm cannot pay is $[K - V_T]^+$, the put option exercise value, and the liability will not be paid in full if this option is in the money. A cheap way to complete the model is to suppose that the firm value follows a

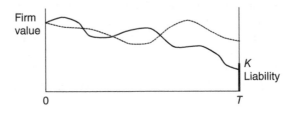

22. **Evolution of firm value.**

log-normal diffusion—the BS model—in which case the default probability is just $\mathcal{N}(d_2)$.

Firm value models cannot directly be used in practice in any straightforward way because the value of a firm is not observable data. It is not the same thing as market capitalization (the share price times the number of shares). Neither is it a traded asset, so it would be meaningless to build what is intended to be a risk-neutral firm value model whose growth rate is the riskless rate. Merton's idea is however valuable conceptually in isolating the essence of credit risk.

Structural form models. Practical structural form models drop the firm value interpretation and regard V_t as an abstract factor process. Following a 1976 paper by Black and Cox, default is deemed to occur at the first time t_D at which V_t reaches some low barrier level L as shown in Figure 23 rather than just checking the value at some payment date as in the Merton model of Figure 22.

If we want to use this model for pricing a CDS then the first task is to calibrate it, that is, select model parameters in such a way that model-predicted prices match the current prices of CDS contracts traded in the market. In this case the 'price' means the premium paid by the protection buyer (expressed as an annualized rate). The CDS contract plays out as follows: either there is no default, in which case the buyer pays the seller the premium coupons over the

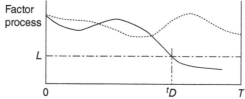

23. Barrier default model, with two trajectories of the factor process, one of which defaults at time t_D.

life of the contract with no payment the other way, or else XYZ defaults at some time before the end of the contract; then the buyer pays coupons only up to the default time and the seller pays the unrecovered post-default value of the XYX bond, that is, the amount by which its post-default market value falls short of the face value. The model is calibrated when the coupon rate is fair in the sense that the expected discounted value of the coupon payments exactly matches the expected discounted value of the post-default payment made by the seller. From the mathematical point of view all these values are expressed in terms of the distribution of the random default time t_D, so the calibration task boils down to specifying a process V_t and a barrier level L such that t_D has a specified distribution. Several ways to accomplish this task have been proposed, and finding an optimal one is still a topic of active research. It can be quite complicated, particularly if default is correlated with the movement of interest rates. How to specify the recovery value is another aspect which has sometimes been contentious, but this is a statistical rather than mathematical question.

To clarify the calibration procedure it might be helpful to consider a highly simplified example, as illustrated in Figure 24. We suppose that two CDS contracts are traded on the same underlying bond of XYZ plc, with maturities $T_1 = 1$ and $T_2 = 2$ years respectively, the protection buyer paying coupons c annually in advance. Thus for the one-year CDS, the buyer pays a single coupon $c = c_1$ at time 0. If XYZ defaults within the first year, the buyer will receive a payment U_1, the unrecovered post-default value of the bond, paid at T_1. For the two-year CDS, a coupon

24. **CDS calibration example.**

payment $c = c_2$ is made at time 0, and a second coupon, also c_2, is paid at T_1 if XYZ has not defaulted before or at that time. The buyer will receive U_1 if default takes place in the first year and U_2, the unrecovered year two value if XYZ defaults in the second year. The coupon rates c_1 and c_2 are market data, set at time 0. We assume that U_1 and U_2 are known constants and that there is no interest-rate volatility, so we also have ZCB values $p_1 = p(0, T_1)$ and $p_2 = p(0, T_2)$.

Our model will specify (albeit indirectly) the risk-neutral distribution of the default time t_D, but it is clear that in this example we just need to know q_1 and q_2, the risk-neutral probabilities of default in the first and second years respectively. These probabilities are fixed by the requirement that for each CDS contract the net discounted value of the cash flows is zero—the contract is a 'fair game'. For the one-year contract, we pay c_1 at 0 and receive U_1 at T_1 if t_D is less than or equal to T_1, which happens with probability q_1. The condition for zero expected discounted value is $0 = c_1 - q_1 p_1 U_1$, fixing q_1 at $q_1 = c_1/(p_1 U_1)$. For the two-year CDS, default happens in the first year, in the second year or not at all, with probabilities q_1, $(1 - q_1)q_2$, $(1 - q_1) \times (1 - q_2)$ respectively. In the first case the cash flows are exactly as in the one-year CDS only with coupon c_2. In the second case we pay two coupons and receive U_2 at T_2; in the third we just pay the two coupons. Writing down the net discounted value condition gives us an equation to solve for q_2; we leave it to the reader to work out the details.

Real cases are much more complicated, in particular because the recovery payments and interest rates will be stochastic and may be correlated with each other and with the default events. But the principle is the same.

Having achieved calibration, our model can be used to price new contracts and to calculate the sensitivity of prices to, for example, movements in interest rates—this kind of information is essential for hedging purposes. It can also be adapted to provide a 'real-world' model for risk management purposes, bringing in statistical data such as the default transition rates of Table 5.

Reduced-form models. We now turn to the second class of default models, those in 'reduced form', which are based on the idea of default intensity: what is the probability that a firm will default tomorrow, given that it is still alive today? This is just conditional probability, governed by Bayes's formula which says that the conditional probability, denoted $\mathbb{P}(H \mid G)$, of an event H given that another event G has occurred is $\mathbb{P}(H \mid G) = \mathbb{P}(H \text{ and } G)/\mathbb{P}(G)$. So suppose we have a random default time T with density function f as shown in Figure 25. The probability that T occurs beyond a given time t is the area under the density function to the right of t, given by $B + C$ in the diagram, so the conditional probability that T occurs in a short interval of time a after t is $B/(B + C)$, and $B \approx f(t)a$. The area A is equal to the distribution function $F(t)$, so if we define $\ell(t) = f(t)/(1 - F(t))$ then $B/(B + C) \approx \ell(t)a$ for small a. We call $\ell(t)$ the *default intensity*. Specifying a default intensity is equivalent to specifying the distribution; in fact, $1 - F(t) = \exp(-\int_0^t \ell(s)ds)$. Note that if F is the exponential distribution $F(t) = 1 - \exp(-ct)$ then $\ell(t) \equiv c$. A reduced form default risk model is ultimately a specification of the default intensity, which can be random. More generally, there may be more than one default time, and other random factors specifying what happens at each time. For example, if we are considering a basket

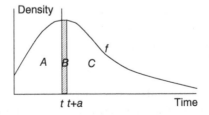

25. **Density of default time distribution.**

of two CDS written on firms XYZ and UVW then we don't know which one would default first, so a reduced-form model would specify the intensity for the first default time T_1 and an indicator variable I equal to 0 if XYZ defaults first and 1 if it is UVW.

Models based on similar principles are widespread in a huge range of applications in statistics and applied probability, any application in fact where the randomness consists of 'point events'—a sequence of random times at which random things happen. A classic example is queueing theory: customers arrive at check-out at random times carrying baskets requiring different amounts of processing time. The arrival rate is analogous to our default intensity. Other applications include insurance (incidence of claims) and medical statistics (incidence of disease, demands on accident and emergency services etc.) as well as, of course, credit risk.

A basic building block in all these applications is the *Poisson process N_t*. Take a sequence of independent random variables S_1, $S_2, \ldots$ each having the exponential distribution with parameter c, so that $\mathbb{P}[S_k > t] = e^{-ct}$, and for $n = 1, 2, \ldots$ let $T_n = S_1 + S_2 + \ldots + S_n$. Think of the T_k as times of 'events', so $S_k = T_k - T_{k-1}$ are the 'inter-arrival' times. For any time t the value of the Poisson process N_t is equal to the number of events that have occurred before or at t, so the sample path is a step-like function as shown in Figure 26. It turns out that, as the name suggests, N_t has the Poisson distribution, with parameter ct. The exponential distribution is

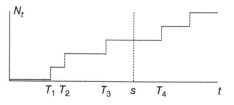

26. Poisson process sample path.

'memoryless': if T is exponentially distributed then
$\mathbb{P}[T > s + t | T > s] = e^{-ct}$: the distribution of future waiting time
doesn't depend on how long we have waited so far. Thus in
Figure 26 the gap between a fixed time s and the next arrival
time (in this case, T_4) has the same distribution as any S_k;
consequently, the process 'restarts' at s, that is, the process $N_t -$
$N_s, t \geq s$ is just another Poisson process, independent of what has
happened before s. The expected value of N_t is $\mathbb{E}[N_t] = ct$, so $M_t =$
$N_t - ct$ has expectation zero and by the memoryless property is a
martingale, $\mathbb{E}[M_{t+s} | \mathcal{F}_t] = M_t$, where $\mathcal{F}_t$ represents the past up to t.

The 'compensated Poisson process' M_t plays a similar role in
modelling point events as Brownian motion W_t plays in relation
to continuous random trajectories: both are independent-
increment processes and can be used as 'inputs' to construct more
complex models. In particular it is straightforward to replace the
Poisson 'rate' parameter c by a stochastic process $\ell(t)$ depending
on other random factors, and to create random 'events' associated
with each jump time T_k. This covers the application to multiple
CDS contracts considered earlier.

A simple example is the 'diamond default model' illustrated in
Figure 27, applicable to modelling possible defaults of two firms.
There are four parameters $c_1, \ldots, c_4$. At the start, the default
intensities of XYZ and UVW are c_1 and c_2 respectively; the
intensity for the first default time is $c_1 + c_2$ and the probability that
XYZ defaults first is $c_1/(c_1 + c_2)$. If XYX defaults first (an upward

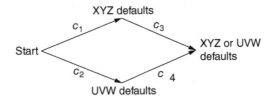

27. Diamond default model.

move, in the figure) then subsequently the intensity for the survivor UVW is c_3. This gives us the possibility to model the effect on UVW of default of XYZ. Similarly, if we have a downward move then UVW defaults first and XYZ then has default intensity c_4. If the firms are totally unrelated then we can set $c_3 = c_2$ and $c_4 = c_1$ so that the default intensities for XYZ and UVW are c_1 and c_2 whether or not the other has defaulted. $c_3 > c_2$ implies that XYZ's default has a negative impact on UVW, while if $c_3 < c_2$ the impact is positive—possibly UVW can take over some of XYZ's market share. This simple model is easily generalized in various ways. For example, there is no reason why the default intensities have to be constant; they could equally well—at greater computational cost—be stochastic. One could, for example, allow the default rates to depend on the current market CDS rates for the two firms.

This example points to a pervasive problem in credit risk modelling: while some parameters can be backed out by the calibration process, there are usually others about which the available data is insufficient for us to do anything more than take an educated guess. How do we know what the effect of default by XYZ will be on UVW, an event that has never hitherto occurred? Only by looking at analogous defaults in the past and/or by examining the business model of XYX and UVW in some detail, giving at best a rough estimate, and this will be a 'real world' estimate. A risk-neutral estimate, needed for pricing, can only depend on market price data for traded contracts, and often there is no relevant data available. The sparsity of data means that credit risk modelling will never be an exact science on a level with the option pricing or interest rate analysis presented in earlier chapters.

A further application of models such as the diamond default model is in connection with Credit Value Adjustments (CVA). In recent years, since the 2008 crisis, financial regulators have imposed the requirement that pricing of essentially all contracts must take into account the possibility of default of the counterparty. At a stroke, this hugely enlarges the amount of quant effort that goes into credit risk evaluation. CVA is the difference between the default-free value of a contract and the value allowing for counterparty default or, if you like, the 'adjustment' that must be made to the 'traditional' pricing method which ignores questions of default. Prior to 2008, this risk was thought to be negligible for trading between banks.

To get some idea what CVA entails, let us consider a plain vanilla interest rate swap, as discussed in Chapter 3, in which at semi-annual coupon dates over N years one party (the payer, P) pays a fixed-rate coupon K while the counterparty (the receiver, R) pays LIBOR L. K is set at time 0 as the swap rate so that by definition the swap has zero value at inception. It also has zero value at maturity, after the final exchange of coupons, as there are no more cash flows. However, between those times it generally has non-zero value. A rise in the swap rate to $K' > K$ favours P, as he could at no cost enter a new swap as receiver; then the LIBOR payments cancel out and he is receiving coupons at a positive net rate $K' - K$ for the rest of the deal. Similarly, a fall in the swap rate is beneficial to R. Now consider CVA from P's point of view. If R defaults at, say, the k th coupon date then P will stop paying coupons too, so P has lost the opportunity to do the reverse trade and receive $K' - K$ net. If $K' > K$ the value of that loss is positive, equal to the sum of the discounted coupon values. But if $K' < K$ then P actually gains on default, since effectively the deal has been cancelled when it is not in his favour anyway. The two cases are sometimes called 'wrong-way risk' and 'right-way risk'. Nowadays, in addition to CVA there are all kinds of other adjustments, collectively known as XVA (plug in your favourite variety for X), which between them are intended to provide tighter risk control. Be that as it may, these adjustments have hugely increased the costs of the firms that have to compute them.

Multi-asset credit risk

We now turn to credit analysis for contracts involving multiple (dozens, thousands) of interest-paying securities. An essential problem arises here. Suppose cash amounts are due to us at various times from different parties (known in this context as 'obligors'). If the obligors have credit ratings then we may very well be able to assess accurately the default probability for each obligor separately. But what about correlation? For an n-vector of random variables there are $n(n-1)/2$ correlation coefficients, so 45 for $n = 10$ or 1,225 for $n = 50$. There is little or no data from which this massive number of correlations can be estimated, so we have to resort to *ad hoc* methods; we will describe two.

The first was developed by RiskMetrics, now a subsidiary of the risk analytics firm MSCI Inc. It relates to firms which both have a credit rating and are listed on a stock exchange. For simplicity, consider just two such firms and assume the rating categories are A, B, C, D (D for default). It is assumed that a rating transition matrix similar to Table 5 is available. Firms I and II are initially rated B and C respectively. Now let X_1, X_2 be a pair of random variables each having standard normal distribution $N(0, 1)$ and correlation r. Referring to Figure 28, divide the x-axis into intervals D, C, B, A whose $N(0, 1)$ probabilities are given by the B row of the rating transition matrix. Do the same for the y-axis using the C row instead of the B row. Taking a sample of (X_1, X_2), the square in which it falls determines rating transitions for firms I, II, each consistent with the probabilities in the rating transition matrix but with a *joint* distribution that depends on the assumed value of r. This value is taken as the sample correlation of the returns of the two firms' stock prices over a one- or three-year period. The procedure is artificial as there is no hard-and-fast reason why this correlation should coincide with default correlation; high stock correlation is, however, evidence that the two firms' fortunes are linked (positively or negatively), so at least this procedure biases the answer in the appropriate direction.

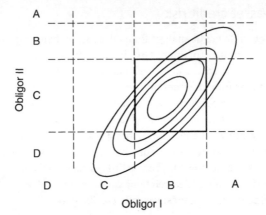

28. CreditMetrics bivariate rating transition model.

Note that to compute the new default probabilities we have to integrate the multivariate normal density over each of the cells determining a specific configuration of new credit ratings. With a large number of obligors this is a huge computation, perhaps best carried out by sophisticated Monte Carlo techniques.

A much simpler procedure was developed by Moody's based on the idea of a *diversity score*. It is known as Moody's BET (Binomial Expansion Technique) and is applicable when all firms in a bond portfolio have similar credit ratings. Each firm is assigned to one of thirty-two industry categories (aerospace, energy, consumer packaged goods . . .). We count the number of instances where a firm is the only one in the portfolio in its industry sector, and the number of pairs, triples, etc. The diversity score defined by Moody's is 1.0 for a single, 1.5 for a double, 2.0 for a triple, and so on for higher groupings. So if we had a portfolio of size ten consisting of three singles, two doubles, and one triple, its diversity score would be $3 \times 1.0 + 2 \times 1.5 + 1 \times 2.0 = 8.0$. Greater concentration means less diversity; the diversity score is equal to the size of the original portfolio only when each member is the unique representative of its industry sector. Now, for the purposes of default risk, we replace the original portfolio of ten bonds by a hypothetical portfolio of eight bonds each having a face value 1/8 of the total face value of the original portfolio. Defaults of the eight bonds are all independent with default intensity equal to some weighted average of the intensities for the original bonds. It follows that if in some period the implied default probability is q the probability that k out of the $n = 8$ hypothetical bonds will default is given by the binomial distribution $(n!/k!(n - k)!)q^k(1 - q)^{n-k}$. This means that we only need to model *one* default intensity process to evaluate the performance of the whole portfolio. Rough and ready? Yes, but necessarily so. In any mathematical modelling task, a more complex model invariably means more parameters. Unless these parameters can be accurately and reliably estimated, extra complexity can easily make a model worse rather than better. Nowhere is this truer than in credit risk where, as alluded to earlier, data is lacking in so many key areas.

Collateralized Debt Obligations (CDO). A primary application of multi-asset credit risk models is pricing and risk analysis of CDOs. These are quite complex structures whose purpose is to stream the income from some portfolio of coupon-paying securities so as to match the appetite of potential investors. It sounds harmless, but unless properly managed CDOs can cause serious damage, the clearest example being the RMBS (residential mortgage-backed security) market in the USA which kicked off the 2008 financial crisis. This is discussed in Chapter 8.

The conventional structure of a CDO is shown in Figure 29. A 'special purpose vehicle' (SPV) is formed for this one transaction, that is, a registered company which will be legally separate from the bank setting up the deal. Ownership of a portfolio of securities ('bonds') is transferred to the SPV, which receives the coupon payments, represented by the arrows in the figure. The SPV now issues its own bonds in several different

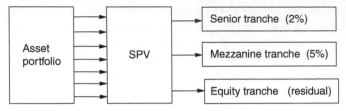

29. Collateralized debt obligation structure.

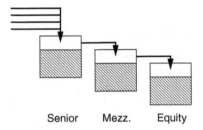

Senior   Mezz.   Equity

30. Cash flow waterfall.

'tranches' (three in the figure, labelled Senior, Mezzanine, and Equity, there could be more), all having the same coupon dates and all but the Equity tranche paying fixed coupons at contractually specified rates, as indicated in Figure 29. The funds needed to pay the coupons are derived from the receipts of coupon payments from the asset portfolio, and these are passed through the 'waterfall' structure illustrated in Figure 30.

Between the coupon dates of the bonds on the right in Figure 29, receipts of coupons received from the portfolio on the left are stored by the SPV. At a coupon date, the cash is streamed to the Senior bond-holders until the coupon due to them is fully paid, so their bucket is full. Then the cash flows through to the mezzanine bucket until that is full, then any remaining cash is paid to the Equity tranche. Thus the Equity tranche has no fixed coupon, it is just paid 'residual receipts', so an investment in it is more like investing in stock than in a bond, since no income is guaranteed; hence the name.

Given the portfolio of assets, designing a CDO means deciding on the number of tranches and, for each tranche, its size (i.e. the total face value of bonds issued) and the coupon it will pay. Constraints are the need to secure from rating agencies a credit rating for each tranche that will make that tranche attractive to investors. It is a lengthy process which doesn't always succeed.

Commentators sometimes deride CDOs as 'alchemy' as the assets in the portfolio, which may be rated no higher than B, are turned into AAA-rated bonds. This seems unfair. Returning to the waterfall analogy, if my house receives the water supply first and my neighbour's house only receives water when my needs are satisfied, then quite clearly I am in a better position than my neighbour, as she will be the first to suffer any shortfall. No base metal has been turned into gold. A more serious criticism is the hand-in-glove relationship between CDO issuers and rating agencies. The latter are paid by the former to carry out the rating process, and may lose market share to less scrupulous agencies if their standards are too strict. Some further discussion of this point will be found in Chapter 8.

Chapter 6
Fund management

Fund management is a huge industry. A survey by the UK Investment Association estimates that assets managed in the UK by its members in late 2016 totalled £6.9tn, with assets managed in the UK on behalf of UK investors rising to £1tn. A scientific basis for fund management was initiated in 1952 when Harry Markowitz—subsequently, an Economics Nobel Prize laureate—published his paper, 'Portfolio Selection', which begins as follows:

> The process of selecting a portfolio may be divided into two stages. The first stage starts with observation and experience and ends with beliefs about the future performances of available securities. The second stage starts with the relevant beliefs about future performances and ends with the choice of portfolio. This paper is concerned with the second stage.

In fact, the first stage is far harder, and there is never a definitive solution. Predicting the future is difficult in most situations, but particularly so for asset prices in financial markets, which are notoriously erratic. Indeed, the 'efficient markets hypothesis', accepted by some economists, postulates that all information relevant to systematic prediction is already 'priced in' to today's market, leaving only 'white noise' as the future. Traders, on the other hand, seek to take advantage of market anomalies to make arbitrage profits, an enterprise that would be impossible in an efficient market.

The difficulties of the first stage do not mean that nothing can be done in relation to the second stage. The beliefs about future performances will include acceptance that the future is grossly uncertain, dealt with by building mathematical models with an adequate degree of volatility, and then well-developed principles can be applied to come up with strategies that mitigate the risks of unfavourable scenarios and take advantage of favourable ones.

One of the principles is diversification, that is, 'don't put all your eggs in one basket'. Suppose we have £1 to invest in two securities at price £1 each, which will produce returns X and Y respectively, where X, Y have the same mean $m > 1$ and standard deviation $\sigma = 1$, and are correlated with correlation coefficient r. We decide to invest £a in the first security and £$(1 - a)$ in the second, giving a 'portfolio' $Z = aX + (1 - a)Y$.

Whatever a, Z has mean m. However the standard deviation of Z does depend on a, as shown in Figure 31, in which a appears along the horizontal axis and the portfolio standard deviation along the vertical axis. The different lines correspond to different values of r. If $r = +1$ then $X = Y$ and there is no 'portfolio effect'. However, if $r < 1$ then the variance is reduced, dramatically so for $r = -1$ when a 50/50 mixture of X and Y has variance 0, giving a sure profit of $m - 1$. We see that combining offsetting risks is a good way to reduce the standard deviation and hence the probability of making a loss.

Markowitz's conceptual contribution was to identify *mean* with *reward* and *variance* (or standard deviation) with *risk*. The latter is not completely straightforward because although the probability of loss increases with variance, we cannot precisely quantify this without knowing more about the return distribution. The controversy here is similar to that surrounding Value-at-Risk

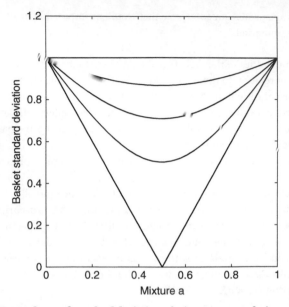

31. Dependence of standard deviation of mixture on correlation. Correlations are 1, 0.5, 0, −0.5, −1 (from top).

(VaR) as discussed in Chapter 7. The advantage is that we now have a problem with only two parameters, the portfolio mean m_p and variance σ_p^2. We define an *efficient portfolio* as one that maximizes m_p subject to $\sigma_p^2 \leq c$ for some specified maximum level c. It is clear that no portfolio that is not efficient can be optimal in this framework.

Optimization based on the Markowitz approach has been surprisingly long-lived in the world of 'traditional' fund management. In a universe of n securities it requires estimation of an n-dimensional mean vector and an $n \times n$ covariance matrix. The mean vector in particular is hard to estimate by purely statistical means and other, more qualitative, input can be used as well. F. Black (of the BS formula) and R. Litterman devised an algorithm (based on Kalman Filtering, for those familiar with it) for blending these two sources of information, which is widely used up to the present time.

One major shortcoming of the Markowitz approach is that it solves a 'one-shot' problem: we invest on Day 1, receive the payoff on Day 2, and that is the end. We could just repeat the process on Day 2, but that could require excessive trading if repeated every day, and there is no concept of long-run optimality. In reality, investment is a process that lasts over time, with revision of the portfolio when deemed appropriate. 'Dynamic' portfolio theory studies strategies in this situation. Aside from the extra realism, it enables us to study a variety of problems. Just maximizing wealth is not the only possible objective for an investor; there are also questions of sustainable spending. If you are managing, say, the London Mathematical Society's endowment fund then your objective is to disburse as much money as possible to support the mathematics community while maintaining the value of the fund so that the Society's activities can continue in perpetuity. There is a standard formula, the 'Yale formula', which says that the Society should invest for maximum growth while disbursing 4 per cent of the capital value of the fund each year. Is that optimal? In other cases, hedging as well as investment is involved. Insurance companies have obligations to make payments as specified by pension and life insurance contracts they have signed. They engage in asset/liability management (ALM) to ensure that they are in a position to meet these obligations as well as staying solvent and providing a return to shareholders.

Mathematical study of dynamic strategies started in the 1950s with a highly stylized, but suggestive, paper by John Kelly of Bell Laboratories concerning repeated independent bets with favourable odds (so that the bettor expects to accumulate wealth). The rules are simple: at each turn the bettor announces his stake X, then a coin is flipped with heads probability $q > 1/2$; the bettor wins X on heads and loses X on tails. Suppose the bettor starts with an initial amount V_0, and express his fortune after N bets as $V_N = V_0 \times 2^{NG_N}$. Then G_N represents the achieved 'growth rate' of the fortune up to time N. For example, if $G_N = 1$ then the bettor has on average doubled his fortune on every turn. This is a standard way to report investment performance: annual reports will record that a fund has grown at 5 per cent over the past year, 3.5 per cent over the last two years (meaning 3.5% per year), etc. In Kelly's problem the bettor's objective is to *maximize the long-run growth rate*. To do this, he chooses to stake a fixed fraction f of his current fortune on each turn, so that at each turn his fortune is multiplied by $(1 + f)$ on a win and by $(1 - f)$ on a loss, so $V_N = V_0(1 + f)^W (1 - f)^L$ where W, L are the numbers of wins and losses up to time N. The corresponding growth rate G_N is

$$G_N = \frac{W}{N} \log_2(1 + f) + \frac{L}{N} \log_2(1 - f).$$

Now we appeal to the *law of large numbers* (see Chapter 2): the sample averages W/N and L/N converge to the corresponding mean values, which are q and $p = 1 - q$ respectively, so the long-run average growth rate is $G = q \log_2(1 + f) + p \log_2(1 - f)$. This is maximized when $f = 2q - 1$, giving a maximum growth rate of $G^* = 1 + q \log_2 q + p \log_2 p$, and this cannot be improved by employing different strategies such as varying f at each turn. The main 'output' from this example is that it is optimal to divide the cake into fractions and invest each fraction differently (in this case, staking it or leaving it on the table). This insight extends to many problems of greater complexity and realism.

In 1969, Robert Merton published a major paper on optimal investment in the context of the BS price model: there are n risky assets (stocks) each of which is modelled by BS with given incremental mean and variance parameters, and the n Brownian motions associated with these assets are all correlated, so we end up having to specify (as in Markowitz's model, but with a different interpretation) an n-vector of means and an $n \times n$ covariance matrix. There is also a 'riskless asset' (cash), a bank account paying a constant rate of interest.

It is quite possible to study optimal growth strategies in this framework, but Merton took a different tack, maximizing 'expected utility' instead, over a fixed time horizon or in the long run. Consumption was also included, so the investor's objective is not necessarily just to get rich. The idea of expected utility goes back to 1738 when it was proposed by Daniel Bernoulli in connection with the so-called St Petersburg paradox, a gambling conundrum of the day. The idea is that investors are 'risk-averse' and their degree of risk aversion depends on wealth. Otherwise put, the marginal value of an extra pound is a decreasing function of wealth, so a graph of value $U(w)$ against wealth w always has positive slope (more is preferred to less) but the slope decreases with increasing wealth; examples are $U(w) = \log w$, $\sqrt{w}$, $-1/w$ or indeed any power w^p/p with $p < 1$, $p \neq 0$. A function with these properties is a *utility function*, and the investor will seek a portfolio with random payoff W that maximizes $\mathbb{E}[U(W)]$. A consequence is that a risky investment will never be preferred to a sure payment unless its expected return is greater than the sure payment. A *risk-neutral* investor is one whose utility function is linear (has constant slope). Investors in this category compare investments only by comparing their expected payoffs. The risk-neutral distribution introduced in Chapter 3 prices everything by expectation, so these prices are 'fair' from the point of view of a risk-neutral investor.

Returning to Merton's problem, the investor starts with wealth $x > 0$ and engages in trading over a time interval $[0, T]$ without, in the simplest setting, extracting anything for consumption or adding any external funds into the portfolio along the way. Otherwise put, the investor is managing a *self-financing portfolio* in which purchases of assets must be funded by sales of other assets. Trading is 'frictionless', meaning that stock can be bought and sold at the same price in arbitrary amounts. The objective is to maximize the expected utility $\mathbb{E}[U(V_T)]$, where V_T is the final portfolio value at time T.

It is a remarkable fact that, for any of the utility functions mentioned, the optimal strategy is to keep *fixed fractions* $f_1, \ldots, f_n$ of portfolio value V_t in each of the stocks, with the remaining value $f_0 V_t$ in cash, where $f_0 = (1 - f_1 - \ldots - f_n)$ in cash. The fractions $f_0, \ldots, f_n$ do not have to be positive or less than 1: short-selling and leverage are allowed. The solution looks superficially very similar to that of Kelly's problem, and indeed the problems are closely related in the case $U(w) = \log w$. The investor will leave all the money in cash ($f_0 = 1$) only when all the stocks have the same mean growth rate as cash—then investing in a stock is taking on risk without any expected reward. Otherwise, there will be long or short positions in all stocks whose mean growth rate is greater or less than the cash rate. In practice, taking a short position is a somewhat cumbersome operation: the stock is borrowed from someone who owns it, and immediately sold for, say, £100. At some later time the stock is repurchased in the market and delivered back to the original owner. If the price has fallen to £90 then the investor makes a £10 profit, so in economic terms is 'short' the stock. Things are so much simpler in futures markets, where the mechanisms for going long or short are exactly the same.

The solution of Merton's problem is easily described but not so easily implemented: because the stock prices are moving around all the time, incessant rebalancing is required to keep the proportions constant. To get something more realistic, one has to take account of trading costs (the bid-ask spread). The analysis is then much more complicated but, in the case $n = 1$ of a single stock, easily described: instead of keeping a fixed fraction f_1 in stock, it is optimal to perform minimal trading in order to keep this fraction between limits f_1^- and f_1^+ that bracket f_1. No trading takes place while the observed fractions lie strictly inside these limits. This kind of analysis is helpful in enabling traders to gauge how far away from the ideal proportions the portfolio has to be before rebalancing is worthwhile.

Following Merton's early work there has been a huge amount of research into dynamic portfolio management, broadening out the range of models that can be considered and providing a general theory. This work has however had nothing like the impact on the practical world of investment that the BS theory and subsequent developments had on the world of option trading, which progressed hand-in-hand with the theory. The reason is easy to see: the models are too stylized and the results depend on parameters that are hard if not impossible to estimate. The very fact that there are so many fund managers in the industry, each with her own investment 'style' indicates that there is no consensus as to optimal behaviour. Indeed, markets would look very different if there were. There is one area in which theory and practice are coming together, and that is short-term trading ('optimal execution') algorithms employed in modern electronic trading platforms. This is discussed in Chapter 8.

There have been many attempts at constructing mathematical models for asset allocation that match real market behaviour more closely. As noted earlier, the basic problem is that markets appear so erratic. Is there anything about them that is more invariant, that is, always the same, or evolves slowly in predictable ways? A very remarkable one was discovered by Robert Fernholz and is shown in Figure 32.

Take all the firms listed on the New York Stock Exchange (at the moment, around 3,000), and arrange them in decreasing order of market cap (capitalization = number of shares × share price). Now divide each one by the sum to give the market cap in proportional terms, and plot a graph on a log-log scale, where the x-axis is the rank and the y-axis the relative market cap. This is Figure 32 where the curves relate to various dates in history from the 1920s to the present day. It is astonishing: the market share of the top

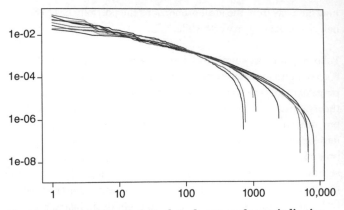

32. Distribution of New York Stock Exchange market capitalization.

hundred or so companies has varied hardly at all in getting on for a hundred years, and the only reason the curves spread out at the small-cap end is that the number of firms listed has significantly grown over time. They all drop off in very much the same way and the top, say, ten firms *never* have more than about 15 per cent of the total market cap.

None of the standard models for multiple stock prices has this feature. In the multidimensional BS model, for example, there is a significant probability that, at some time, a small group of stocks will constitute a large fraction of the total market cap, so clearly the model is missing something. It has been argued that there are significant differences in growth rates in different bands of market cap, and that investment strategies could be devised to take advantage of this feature. To do so, we need a model that does take rank into account, in particular influencing the growth rate. This is, however, far from straightforward because it means that the growth rates of stocks A and B have to change whenever A overtakes B in rank or vice versa. This has led to a whole new mathematical theory of 'rank-dependent diffusions', one of the many instances in which questions arising in finance have led to new research in probability.

The asset allocation problems described so far relate to investment horizons of at most a few years. What are we going to do about the long-term hedging problems faced in ALM in areas such as pensions and life insurance, where the liabilities can stretch out fifty years or more into the future? Prediction on this scale really is 'Knightian uncertainty': as Keynes put it, there is 'no scientific basis' for any probabilistic model. But business must go on, as indeed it has for hundreds of years, but the increasing speed and complexity of modern finance demands new and improved methods. The basic idea is to concentrate more on avoiding disaster than maximizing profits. Lacking credible dynamic models for the future, one falls back on the entirely defensible approach of 'what if?' analysis. A repertoire of possible future 'scenarios' is drawn up with a view to devising strategies such that the firm stands a high chance of survival even in the most unfavourable circumstances.

One way to represent the scenarios is in the form of a 'scenario tree' as shown in Figure 33. (This figure is purely for illustration—a real one would be far bigger.) Time 0 is the current time, and 1,2,3 represent times in the future, unevenly spaced; the gap 0–1 might be six months or a year and the gaps get increasingly long so that in total the time covered is, say, thirty years. The idea is of course that the future becomes more vague the further away it is. For the same reason, the number of branches from a node is large at node zero, where we have accurate information, and reduces further out. In Figure 33 we have a total of thirty scenarios: every possible path from left to right. At each node, a list of the values of various economic variables, stock index levels, interest rates, bond yields, economic growth figures, etc., is specified, along with liabilities that must be paid. Probabilities of the possible branches from each node—twenty-one nodes in the

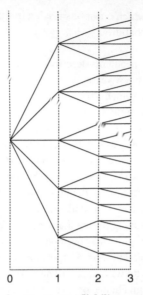

33. Scenario tree for long-term asset/liability management.

figure—are specified in as plausible a manner as possible.
At each node we specify a penalty for failure to meet the
liabilities. We now have a mathematical optimization problem:
maximize the expected profit minus penalties over the possible
asset allocations. It turns out that the solution is not very sensitive
to the assignment of probabilities; if the penalties are sufficiently
severe then only strategies that largely avoid the states where
penalties are incurred can be optimal, even if these states have low
probability. The main determining factor is a judicious choice of
scenarios.

When the optimal solution has been computed, positions are
taken in the market corresponding to the time-0 optimal
allocations. The remaining decisions are never used. When the
time comes to review asset allocations, the whole process is

repeated, with the review time becoming 'time 0'. The scenario tree approach has proved highly successful is a variety of applications. Of course it is as much an art as a science, but that is inevitable when the time horizon is much longer than the maturity time of contracts available in the market.

Chapter 7
Risk management

The risk management function of a financial company monitors a whole zoo of risks that the company faces: market risk, credit risk, liquidity risk, operational risk (the risk of internal errors such as 'fat finger syndrome'), reputational risk, legal risk, etc., etc. Some of these are connected to regulatory requirements, while others are internal procedures designed to assist the management of the company's assets and liabilities. It is a big subject, and in this small chapter we focus exclusively on those areas in which mathematical finance plays an important role. Credit risk was discussed in Chapter 5; here we will discuss market risk, which is concerned with assessing how sensitive the value of the company's trading book is to anticipated movements in the market prices of the assets it contains. Evaluations are carried out at various levels of aggregation from individual trading desk to the company as a whole.

Market risk management is a curious mixture of real-world and risk neutral probabilities. Suppose we hold a call option on some underlying asset which matures six months from now, and its value now is £100. What might it be worth one week from now? Obtaining a value is a two-stage process: first we need to predict the value of the underlying asset and the implied volatility of the option one week ahead. That is a real-world calculation. But having made these predictions, the option value is given by the BS formula with those predictions plugged in, a *risk-neutral* expectation. One hardly notices this when the option is priced by a formula, but it becomes a live issue when the whole calculation is done by statistical simulation: then one needs to generate random price paths of the underlying asset following the real-world distribution for a week and then switching to the risk-neutral distribution.

In this context a point forecast is not very useful, it is more pertinent to compute the (real-world) distribution of the value one week ahead. But this is too complicated and what is needed is some carefully chosen statistic that summarizes the risk profile in a straightforward but informative way. In previous chapters, particularly in connection with optimal investment, the conventional representation of risk has been standard deviation. However, this mixes up gains and losses, while the latter are clearly more relevant from the 'risk' point of view. Quantiles are more informative, the y-quantile q_y of a distribution being the value at which the distribution function crosses level y, so the probability that the corresponding random variable is less than q_y is equal to y. See Figure 34.

Estimating quantiles and later observing the actual outcomes provides a highly intuitive and informative way to tell whether a risk-management algorithm is well calibrated. A typical result is shown in Figure 35, where the x-axis is time marked in (business) days and the y-axis is portfolio value. At each day the solid vertical

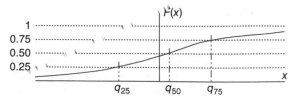

34. **Twenty-five, 50, and 75 per cent quantiles of distribution function F.**

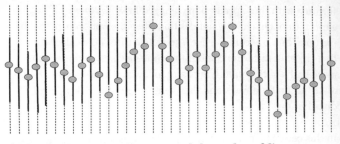

35. Predicted inter-quantile ranges and observed portfolio returns.

line represents the interval between the 5 per cent and 95 per cent quantiles of the portfolio return distribution (the inter-quantile range) that was predicted one week earlier, and the grey circle is the actual return over the past week. Since the inter-quantile range has probability 90 per cent, we expect that 10 per cent of the returns will be outside the range, and in the figure this is roughly true. Also the true returns are sprinkled quite uniformly over the range. If they had all been much closer to the centre of the range that would be evidence that the prediction model is mis-calibrated—the predicted range should be smaller. One can see at a glance whether the prediction model is doing its job.

The monitoring procedure just described is attractive but has one defect: implicitly it assumes the future will be much like the recent past. There is no allowance for occasional sudden 'shocks' that would trigger major losses. Separate measures, specifically taking care with the portfolio composition so that overall it is well hedged, must be taken to guard against shocks.

Predicting the return quantiles can be done in various ways. One of them is to use an econometric model, as outlined in Chapter 2, with parameters estimated from data from the recent past and/or from various 'stressed' periods in history if one wants a more conservative estimate. Having calibrated the model, predictions can be calculated either analytically or via Monte Carlo simulation. Alternatively, a more data-driven approach can be used by looking at empirical distributions of past returns.

Turning to regulatory requirements imposed on the banking sector, these are determined internationally by the 'bankers bank' BIS (the Bank for International Settlements) in Basel, otherwise known as the 'Basel Committee'. It promulgates rules requiring banks to hold a buffer of capital related to the perceived riskiness of their trading positions. In contrast to the monitoring procedure described earlier, these rules are entirely concerned with ensuring that banks avoid life-threatening losses so, referring to Figure 34, they refer exclusively to the extreme left-hand end of the return distribution. The question is what 'risk measure', that is, statistic, the bank should be required to compute. For a long time after its introduction in the 1990s the answer to that question was 'VaR', a quantity measured in monetary terms, say dollars. Figure 36 shows the definition.

The function F is the distribution of X, the change in portfolio dollar value (not return) over some given holding period. Thus the portfolio has made a profit if $X > 0$ and a loss if $X < 0$. The VaR at level a, denoted VaR_a, is the value x such that $F(-x) = 1 - a$. Here a is a number close to 1, generally written as a percentage, as in $a = 99\%$, so the statement $\mathrm{VaR}_{99\%} = \$1\mathrm{m}$ is saying that with 99 per cent probability the loss will not exceed $1m.

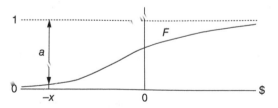

36. Distribution function F of change in portfolio value.

aR is not without controversy. First, how to estimate it at extreme levels like 99 per cent is far from obvious. Unlike the monitoring process of Figure 35, one cannot just observe the exceptions (the points outside the inter-quantile range) and check that they occur with the appropriate frequency: at the extreme negative end, one sincerely hopes *never to observe* an exception. More fundamentally, VaR ignores the magnitude of losses beyond VaR. It could be that some really huge loss is lurking at a probability just slightly less that of the VaR threshold. Surely this possibility should somehow be accounted for.

In a highly influential 1999 paper, Artzner, Delbaen, Eber, and Heath set about formulating axiomatic criteria that a satisfactory risk measure should satisfy. In their formulation, a risk measure attaches a number $R(X)$ in dollars to each random portfolio X, interpreted as a portfolio value. The axioms are

(i) $R(0) = 0$.
(ii) If, with probability 1, $X_1 \leq X_2$ then $R(X_1) \geq R(X_2)$.
(iii) For any constant a and portfolio X, $R(X + a) = R(X) - a$.
(iv) For any two portfolios X_1 and X_2 we have $R(X_1 + X_2) \leq R(X_1) + R(X_2)$.
(v)   For constant $a > 0$ and portfolio X, $R(aX) = aR(X)$.

Axiom (i) just says that if you are out of the market you run no risk; (ii) says that if portfolio X_2 always outperforms X_1 then X_2 is less risky than X_1; (iii) formalizes the idea that risk is measured in cash: if I give you \$1 then your risk is reduced by \$1; (iv) is the 'diversification' axiom: if you combine two portfolios then the risk of the new big portfolio cannot be more than the sum of the individual risks. Finally, (v) encapsulates 'scalability': double the portfolio size, double the risk. A risk measure satisfying all these conditions is said to be *coherent*.

The first four axioms are uncontroversial. Axiom (v) is more debatable as it apparently ignores liquidity questions: a huge portfolio could incur extra risks related to the difficulty of realizing its value in the market, whereas a smaller one could be more liquidly traded. Later contributors have replaced (iv) and (v) by a convexity condition that permits liquidity risk to be included. Adding axiom (v) to this implies coherence, that is, (iv) is then satisfied as well.

Although $R(X)$ is defined as a function of a random variable X, in almost every application it only depends on F_X, the distribution of X, as in the case of VaR, so we can write $R(F)$ for the risk of a portfolio whose distribution is F.

It turns out that VaR is not a coherent risk measure; it is possible to find examples in which the diversification axiom (iv) is not satisfied, although these examples are somewhat artificial and many have argued that (iv) will generally be satisfied for realistic portfolio distributions. Nonetheless, the combination of non-coherence and the loss-beyond-VaR question has lead to a search for better alternatives, and the spotlight is now turned on the so-called Expected Shortfall (ES) risk measure, which is coherent and does incorporate the effect of improbable but large losses. ES is defined as the *conditional expected loss beyond* VaR_a. Figure 37 shows the idea: ES_a is VaR_a plus the value of the 'put option' shown in the figure, conditional on it being in the money, the latter term being equal to $1/(1 - a)$ times the expected value of the put payoff.

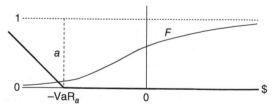

37. Expected Shortfall.

By definition, ES_a is never less than VaR_a. The reader can check that the two are equal only when there is positive probability $(1 - a)$ of a loss of size VaR_a and probability zero of bigger losses—so this is the most optimistic possible way in which losses beyond VaR_a could occur. ES_a allows big losses to spread out over a larger range.

In recent years the BIS regulations are increasingly specifying ES in place of VaR as the approved risk measure, but there has been some backlash. Several counter-arguments have been advanced, but here we will concentrate on one of them: the sheer difficulty of computing that apparently innocent thing, an expectation. An alternative and equivalent definition of ES is the integral

$$ES_a = \frac{1}{1-a} \int_0^{1-a} VaR_s \, ds. \tag{12}$$

To compute this we need to know VaR_s for all levels s between 0 and $1-a$, but no amount of data will give us this information directly, since VaR_s increases to infinity as s approaches 0 and consequently any errors in estimating the large-loss quantiles can lead to significant errors in estimating ES. In summary, the calculation of ES is highly sensitive to things we cannot actually observe. So what can be done? Unsurprisingly, a huge amount of work by statisticians and econometricians has addressed this issue. Some of the methods used are as follows.

(a) Build econometric models such as the GARCH model described in Chapter 2 using input 'noise' variables W_k whose distributions are chosen to reflect the empirical distributions of such data as is available. The model is then calibrated to the portfolio data under consideration, following which the ES can be calculated by Monte Carlo simulation (or, occasionally, analytically). This gives us an answer, but does make the tacit assumption that extrapolation from the observed data is reliable.

(b) Use 'extreme value theory'. This is a branch of statistics, dating back originally to R.A. Fisher and L. Tippett in 1928, on which a huge literature now exists. Recall the CLT described in Chapter 2, which says that if we take a sequence of suitably normalized sums of i.i.d. random variables then the distributions of the terms in the sequence converge to the normal distribution. In the CLT the normalization consists, for the nth term, of subtracting the mean and dividing by $\sqrt{n}$. In extreme value theory, 'sum' is replaced by 'maximum' and we ask the same question. The Fisher–Tippett theorem asserts that if a suitable norming sequence exists then the limiting distribution must belong to one of three categories, each category consisting of a one-parameter family of distributions. Then it is possible to disentangle, for each of the three categories, what properties of the distribution function for the underlying i.i.d. variates lead to an extreme value distribution in that category. This can help us in categorizing the extreme-loss distributions in portfolio models.

(c) A more radical approach is to abandon ES altogether, replacing it by EMS—expected *median* shortfall—in which instead of taking the conditional mean of the shortfall we take the conditional median. But $\text{EMS}_a = \text{VaR}_b$ where $b = (1 + a)/2$, so calculating EMS requires only VaR technology.

In the end, the main point of calculating ES is to estimate the gap between ES and VaR in order to quantify the extra capital buffer that should be required to allow for losses beyond VaR. Since there is so much uncertainty about estimating these small probabilities EMS, just equal to VaR at a higher confidence level, seems a definite contender.

This discussion leads to the next question: having settled on an algorithm for computing a risk measure, how can we tell whether the answers it produces are correct? Nothing can be said about a single calculation in which we produce a value of the risk measure and then eventually observe the portfolio value X that materializes; whatever value of X is observed we cannot say whether the risk measure calculation was correct or not. But with a sequence of such calculations and outcomes we can, as in Figure 35, assess whether the process is well calibrated. Here VaR has a decisive advantage, hinging on a well-known result called the probability integral transform (PIT); if X is a random variable with continuous distribution function F then the random variable $U = F(X)$ has the uniform distribution on the unit interval, that is, it is a 'random number' in the computing sense of the term. Further $X < q_a$ is the same thing as $U < 1 - a$. You can see all this from Figure 34: the probability that X falls in the interval from q_{25} to q_{50} is 0.25, and this is the probability that $U = F(X)$ lies between 0.25 and 0.5. So the probability that U lies in any interval is equal to the length of that interval; that is the uniform distribution. Now suppose that we observe portfolio values X_1, $X_2, \ldots$ over time and, at each time $k - 1$ we forecast q_k the quantile corresponding to to the VaR$_a$ of the next observation X_k. Then we define a random variable B_k to be 1 if X_k is beyond the VaR threshold and 0 if it is not. This is a 'Bernoulli' random variable—it just takes the values 0 or 1—and it follows from the PIT that *if* our calculations are correct the probability that $B_k = 1$ is $(1 - a)$ and that all the B_ks are independent. So correct calculations mean that we have generated an i.i.d. sequence of Bernoulli random variables. There is a whole battery of tests in the statistical literature aimed at accepting or rejecting the i.i.d. hypothesis for a Bernoulli sequence. By using these test we can judge whether our VaR calculation procedure is well-founded or not.

There is no other risk measure for which the possibilities for verification of correctness are so favourable. For risk measures such as ES that involve expectations it is possible to devise criteria that will be satisfied whenever the forecasts are correct, but these are less universal than the PIT test and not amenable to standard statistical tests. This is one good reason for the backlash against ES and in favour of VaR.

From the discussion in this chapter one can see that risk management is a very different game to the 'classical' mathematical finance of preceding chapters. The reason is that we are not directly concerned with pricing, but rather with outcomes. Outcomes occur in the real world, not the risk-neutral world so, naturally, techniques from statistics and econometrics are the way to assess them. At the same time 'valuation' and 'outcomes' are two sides of the same coin and it adds value to approach both sides with a holistic point of view.

Chapter 8
The banking crisis and its aftermath

In this chapter we look at what happened in 2008 and subsequently. The banking crisis was a cataclysmic event that, one way or another, has affected the lives of most people on the planet. It would be out of place to go into these wider implications here. Consonant with the aims of this book, I will discuss two questions in the two sections of the chapter: what happened to the banks; and what has been the fallout for mathematical finance?

The banking crisis of 2008

We are all familiar with grand bank buildings, constructed on an epic scale and endowed with Greek-columned porticos and other evocations of antiquity. A trend-setting example was the Bank of England, designed in the neo-classical style by Sir John Soane, who was appointed in 1788 and worked on the building on and off for the next forty-five years. What is the point of these extravagant constructions? Clearly, it is to project an image of the Bank as a rock of stability, trustworthiness, and reliability in an uncertain world. In reality, nothing could be further from the truth. A bank is an organization that does not produce anything tangible and—thanks to the institution of fractional reserve banking—cannot pay its depositors on demand. Every banker dreads the day when, as happened to Northern Rock (note the name) on 14 September 2007, customers are found queueing at the door asking for their money back. It was the first UK bank run since 1866. There have been far more bank failures in the USA, mainly sma local banks in the depression era. Indeed, Franklin D. Roosevel first 'fireside chat' radio broadcast on 12 March 1933, eight day after taking office, concerned the banking crisis, and the Presic explained to his listeners in words of one syllable that when yo deposit your money in a bank, the bank doesn't just store it in vault.

Because of fractional reserve, the bank is highly 'levered'—its liabilities exceed its equity (i.e. the bank's own money) by a large factor. This means that relatively small changes in the value of its liabilities have a large effect on the bank's net position. To balance things up, banks engage in inter-bank borrowing and lending; there is a well-developed system by which short-term loans (overnight to a few months) can be quickly arranged. In banking jargon, the inter-bank market is highly liquid. LIBOR, the London inter-bank offer rate that we discussed in Chapter 4, is just an average of rates charged by banks for lending to each other. The 2008 crisis was a 'liquidity crisis' caused by the inter-bank market seizing up. The reason for this was dislocation in the sub-prime mortgage market, a topic that is exposed in all its gory detail in Michael Lewis's book *The Big Short* or in the excellent film of the same name.

We discussed CDOs in Chapter 5. Assets are bundled together and the income from them is streamed to bondholders in various categories, the top category being the 'senior tranche', rated AAA with the implication that bonds in this tranche are only fractionally less secure than a government bond. When the assets are mortgages the CDO is called an RMBS so the coupons paid to the SPV are the mortgage payments made by the householders, of whom the would be hundreds in any one RMBS.

It is beyond the scope of this book to go into complete detail here, but in brief the story is this. In the 1990s and 2000s the US

Congress introduced measures to encourage 'affordable housing'. To make this work, mortgage providers had to increase the proportion of sub-prime mortgages issued, meaning more expensive mortgages granted to people whose income and/or credit record would prevent their qualifying for a regular mortgage. The mortgages were then packaged into sub-prime RMBS, providing attractive returns for investors given that the upper tranches were still awarded top credit ratings. These tranches were heavily marketed to investors all over the world. In 2006 America suffered a nationwide house-price slump, causing mayhem in the mortgage market as mortgage holders slipped into negative-equity territory, and the RMBS tranches lost value rapidly. A quirk of US mortgages is that the holder has a 'put option' on her house: she has the legal right to hand in the keys to the bank and walk away. Many did.

The first banking casualty was New York bank Bear Stearns which had launched two RMBS-based hedge funds. On 16 July 2007, Bear Stearns announced that the two funds had lost almost all value, and eventually, in March 2008 after a brief intervention from the New York Federal Reserve Bank (the Fed), Bear Stearns was sold to J P Morgan-Chase for $2 a share. By this time banks all over the world were awash with securities that nobody knew how to value but were certainly not worth much. Trust between banks withered and the inter-bank lending market froze. The most spectacular casualty was Lehman Brothers, whose fortunes spiralled downhill over heavy participation in the RMBS market. In September 2008 it was unable to borrow from other banks, and the Fed declined to step in, leading to the biggest bankruptcy in American corporate history on 15 September. By this time the financial crisis was in full swing, with banks unable to borrow funds commercially because of doubts about their viability, leaving them dependent on government bail-outs.

The story is a toxic mixture of mismanagement, miscalculation, and outright fraud. There is plenty of blame to go round and the quants cannot completely absolve themselves, although arguably the rating agencies should accrue more. It was certainly a mistake to award a triple-A rating to any part of an RMBS. As mentioned earlier, only a handful of real corporations are awarded this gold-star rating, yet we arrived at a situation where thousands of CDO and RMBS tranches had it too. Was it really credible to claim that every one of these tranches was *pari passu* with Johnson & Johnson in terms of default risk?

The procedure for constructing a CDO was outlined in Chapter 5. Many have rightly pointed to the symbiotic relationship between rating agency and trading desk in this process as a prime cause of the errors that were made. But this is a technical point. Ultimately the board of a company is responsible for managing it, and most CEOs proved unable to resist the temptation to dance while the music was playing. They should have examined the state of the dance floor first.

Post-crisis mathematical finance

Since 2008 three factors have been at play, which between them have revolutionized the way the financial sector works. They are: (i) changes in the regulatory environment and other measures designed to prevent the recurrence of banking crises, (ii) rapid development of new technology, leading to a whole new industry under the umbrella name of FinTech, and (iii) the long-term persistence of ultra-low interest rates, causing a major shift in the balance between the present and the future. Item (iii) is beyond the scope of this book, but items (i) and (ii) have greatly affected the aims and scope of mathematical finance in ways that we now describe.

In response to the crisis the regulatory framework of the banking sector has been significantly tightened up in a bid to ensure that banks do not run out of money. What this means from the quant perspective is that many quants have been shifted from revenue-generating trading desks to the cost centre of risk management. There, their job is to compute the long list of value adjustments, or 'XVA' as outlined in Chapter 5. One effect of this change has been to kill any market for 'exotic' options with elaborate payoffs: with the adjustments in place, even a plain vanilla call option is 'exotic'. Instead, the emphasis is now on efficient trading.

The crisis was caused by inter-bank lending freezing up, and in response this a new research area has emerged concerned with the stability of banking networks. The work followed on from a highly influential 2001 paper by Eisenberg and Noe, who modelled the settlement process, for a group of banks each of which has assets and may have pending contracts to pay or receive money from other banks. Then someone fires the starting gun and these obligations must be settled. A bank will fail if its net liabilities exceed its assets, and the point is that this can happen indirectly: Bank A can fail because Bank B does not make the promised payment, and that has happened because C failed to pay B. The 2001 paper formulates conditions under which an orderly settlement takes place. Since then, further studies have investigated conditions under which failure at one node of the network will propagate over time to nodes at some distance away, and hence what are the characteristics of 'safe' networks. This points to another characteristic of post-crisis mathematical finance. Network theory is a well-established discipline with applications from transportation to biology, and in general workers in mathematical finance now find themselves in contact with a far wider range of other scientists than used to be the case.

Turning to trading, we saw in Chapter 3 that *volatility* is the key component in pricing any contract involving optionality. Research in this area has received a new boost due to the introduction by Gatheral of so-called 'rough volatility'. Recall that to avoid arbitrage, asset price models have to be semi-martingales. However, volatility is not directly a traded asset and it turns out,

120

perhaps surprisingly, that various processes that are not semi-martingales can by very successful in providing models that calibrate easily to market implied volatility surfaces such as the one shown in Figure 15.

A further element of trading concerns price sensitivity. The classic theory ignores bid-ask spreads and assumes that prices are unaffected by trading, in contrast to reality where it is clear that if one attempts to sell a huge quantity of stock all at once then one will not achieve the pre-trade price. Almgren and Chriss in 2001 came up with a simple but highly effective way to deal with this. They assumed that trading takes place at a finite rate and the price impact is a drop in price proportional to the rate of selling. Then they were able to specify the optimal strategy for selling a certain quantity of stock in a fixed time (say, one day). In the intervening years the nature of trading has changed dramatically with the deployment of today's computer technology, specifically (a) the trading frequency has increased by orders of magnitude, so that in some markets prices may change on a microsecond time scale, (b) the great majority of trades are executed by algorithms of which the Almgren and Chriss algorithm is an early prototype, and (c) more diverse sources of information, obtained for example from social media, are added to the mix. We have seen occasional dislocations of the market in the form of 'flash crashes' where a massive drop in prices suddenly takes place, only to be reversed in twenty minutes. Such phenomena must be related to the interaction of competing algorithms, but the exact mechanisms are unclear and the whole subject surely demands more research. Investigators might also consider the social benefits of ultra-fast trading: is faster better in some well-defined way?

One approach to the study of trading is to examine the 'market microstructure', the actual price-formation mechanism. In many markets this is a LOB, illustrated schematically in Figure 38. Participants can post a buy (bid side) or sell (offer side) order either as a 'limit' or 'market' order. In the former case a price is

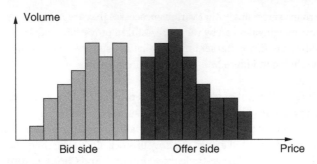

Volume

Bid side   Offer side   Price

38. Limit order book.

stated, and the order book is a record of unexecuted limit orders.
The bars in the figure represent the number of units of the asset
for which orders have been placed at that price (prices being
quantized to multiples of the 'tick size', typically 1 cent in US
markets). The two sides never overlap because if a bid comes in
that is higher than an offer-side price then that order will be
executed immediately. Market orders are always executed
immediately, at the best available price, while limit orders are just
added to the book. The gap between the best offer and bid prices is
the *bid-ask spread*. Limit orders can also be cancelled. We can get
some idea of price impact by observing the shape of the order
book. A market sell order will be executed at the highest bid price
until the inventory at that price (the rightmost bar on the bid side
of the chart) is eaten up, and then the order moves on to the next
available entry at a lower price.

There is obviously scope for dynamic modelling of the evolution of
the book on the basis of which one could derive optimal strategies.
Since movements in the order book correspond to discrete events,
that is, arrival of orders, it seems that the appropriate modelling
framework will include point processes similar to those of
reduced-form credit risk, as discussed in Chapter 5, albeit at a
totally different time scale. Then one could consider a central-limit

type scaling to get a big-picture view. It is certainly a challenge to turn such ideas into reality.

Another possibility for handling the order book and other questions in optimal trading is to use machine learning, an increasingly widely used technique in the world of finance. Crudely put, machine learning is a set of techniques in which incoming data is processed through a network-like structure depending on a finite number of parameters to produce an output which could be, for example, a prediction of future data. A 'training' set of data is used to estimate the parameters. The basis for the network structure goes back to a fundamental theorem in algebra called the Kolmogorov–Arnold representation theorem. This says that any multivariate function f of n variables can be written as a finite composition of continuous functions of a single variable together with addition; specifically, the representation takes the form

$$f(x_1, \ldots, x_n) = \sum_{q=0}^{2n} F\left(\sum_{p=1}^{n} g_{q,p}(x_p)\right).$$

This totally remarkable result justifies attempts to disaggregate a transformation $y = f(x)$ into a concatenation of elementary pieces, which is what machine learning does.

In the case of the LOB, application of machine learning means identifying features that permit reliable prediction of price movements, reliable in the sense that trades resulting from these predictions are profitable most of the time, and development of execution algorithms that capture this profitability with low trading costs.

As can be seen from this quick *tour d'horizon*, current developments require today's quant to assemble a large—larger than in earlier times—'toolkit' of conceptual understanding, mathematical technique, and computational skills to do her job. The challenges are diverse, and not centred around a 'grand challenge' problem such as the FTAP puzzle of the last century. Also, finance is no longer the exclusive property of a few large investment banks, as other actors ranging from hedge funds to cryptocurrency markets take advantage of the new technology—and the technology spills over into non-finance industries such as transportation or health care. It's a great time to be there!

References

Chapter 1: Money, banking, and financial markets

John Maynard Keynes, *The General Theory of Employment, Interest and Money*. Macmillan, 1936

Chapter 2: Quantifying risk

John Maynard Keynes, The general theory of employment, *Quarterly J. Economics* 51 (1937), 209–23

Chapter 3: The classical theory of option pricing

L. Bachelier, Théorie de la speculation, *Ann. Sci. Ecole Norm. Sup.* 17 (1900) 21–86. Reprinted in M. Davis and A. Etheridge, Louis Bachelier's 'Theory of Speculation': the Origins of Modern Finance. Princeton University Press, 2006

F. Black and M. Scholes, *J. Political Economy* 81 (1973) 21–86

Chapter 6: Fund management

H. Markowitz, *J. Finance* 7 (1952) 77–91

Further reading

Chapter 1: Money, banking, and financial markets

P. Samuelson and W. Nordhaus, *Economics*, 19th edn. McGraw-Hill International Edition, 2010

J. C. Hull, *Options, Futures and Other Derivatives*, 10th edn. Pearson, 2017

Chapter 2: Quantifying risk

O. Linton, *Probability, Statistics and Econometrics*. Academic Press, 2017

F. Knight, *Risk, Uncertainty and Profit*. Dover Publications, 2006 (originally published 1921)

Chapter 3: The classical theory of option pricing

M. Baxter and A. Rennie, *Financial Calculus*. Cambridge University Press, 1996

N. H. Bingham and R. Kiesel, *Risk-Neutral Valuation*, 2nd edn. Springer, 2004

B. Øksendal, *Stochastic Differential Equations: An Introduction with Applications*, 5th edn. Springer, 2010

Chapter 4: Interest rates

A. Cairns, *Interest Rate Models*. Princeton University Press, 2004

D. Filipović, *Term Structure Models*. Springer, 2009

Chapter 5: Credit risk

Z. Capiński and T. Zastawniak, *Credit Risk*. Cambridge University Press, 2017

D. Brigo, A. Pallavicini, and R. Torrisetti, *Credit Models and the Crisis*. Wiley, 2010

G. Chacko, A Sjöman, H. Motohashi, and V. Dessain, *Credit Derivatives: A Primer on Credit Risk, Modeling, and Instruments*, 2nd edn. Pearson FT Press, 2015

Chapter 6: Fund management

T. Björk, *Arbitrage Theory in Continuous Time*, 3rd edn. Oxford University Press, 2009

M. Davis and S. Lleo, *Risk-Sensitive Investment Management*. World Scientific, 2015

L. MacLean, E. Thorp, and W. Ziemba, *The Kelly Capital Growth Investment Criterion*. World Scientific, 2011

Chapter 7: Risk management

P. Christoffersen, *Elements of Financial Risk Management*, 2nd edn. Academic Press, 2016

Chapter 8: The banking crisis and its aftermath

J. Gatheral, *The Volatility Surface*. Wiley, 2006

T. Geithner, *Stress Test: Reflections on Financial Crises*. Random House Business, 2015

X. Guo, T.L. Lai, H. Shek, and S. Wong, *Quantitative Trading*. CRC Press, 2017

M. Lewis, *The Big Short*. Penguin Books 2011

Index

Index

Economics
A Very Short Introduction
Partha Dasgupta

Economics has the capacity to offer us deep insights into some of the most formidable problems of life, and offer solutions to them too. Combining a global approach with examples from everyday life, Partha Dasgupta describes the lives of two children who live very different lives in different parts of the world: in the Mid-West USA and in Ethiopia. He compares the obstacles facing them, and the processes that shape their lives, their families, and their futures. He shows how economics uncovers these processes, finds explanations for them, and how it forms policies and solutions.

'An excellent introduction . . . presents mathematical and statistical findings in straightforward prose.'

Financial Times

GAME THEORY
A Very Short Introduction
Ken Binmore

Games are played everywhere: from economics to evolutionary
biology, and from social interactions to online auctions. Game
theory is about how to play such games in a rational way,
and how to maximize their outcomes. Game theory has seen
spectacular successes in evolutionary biology and economics,
and is beginning to revolutionize other disciplines from
psychology to political science. This *Very Short Introduction*
shows how game theory can be understood without mathematical
equations, and reveals that everything from how to play poker
optimally to the sex ratio among bees can be understood by
anyone willing to think seriously about the problem.

www.oup.com/vsi

NUMBERS
A Very Short Introduction
Peter M. Higgins

Numbers are integral to our everyday lives and feature in everything we do. In this *Very Short Introduction* Peter M. Higgins, the renowned mathematics writer unravels the world of numbers; demonstrating its richness, and providing a comprehensive view of the idea of the number. Higgins paints a picture of the number world, considering how the modern number system matured over centuries. Explaining the various number types and showing how they behave, he introduces key concepts such as integers, fractions, real numbers, and imaginary numbers. By approaching the topic in a non-technical way and emphasising the basic principles and interactions of numbers with mathematics and science, Higgins also demonstrates the practical interactions and modern applications, such as encryption of confidential data on the internet.

www.oup.com/vsi